The Human Exploration of Space

Committee on Human Exploration

Space Studies Board
Commission on Physical Sciences, Mathematics, and Applications
National Research Council

NATIONAL ACADEMY PRESS
Washington, D.C. 1997

NOTICE: The projects that are the subject of this volume were approved by the Governing Board of the National Research Council, whose members are drawn from the councils of the National Academy of Sciences, the National Academy of Engineering, and the Institute of Medicine. The members of the committees responsible for the three reports collected and reprinted in this volume were chosen for their special competences and with regard for appropriate balance.

The National Academy of Sciences is a private, nonprofit, self-perpetuating society of distinguished scholars engaged in scientific and engineering research, dedicated to the furtherance of science and technology and to their use for the general welfare. Upon the authority of the charter granted to it by the Congress in 1863, the Academy has a mandate that requires it to advise the federal government on scientific and technical matters. Dr. Bruce Alberts is president of the National Academy of Sciences.

The National Academy of Engineering was established in 1964, under the charter of the National Academy of Sciences, as a parallel organization of outstanding engineers. It is autonomous in its administration and in the selection of its members, sharing with the National Academy of Sciences the responsibility for advising the federal government. The National Academy of Engineering also sponsors engineering programs aimed at meeting national needs, encourages education and research, and recognizes the superior achievements of engineers. Dr. William A. Wulf is president of the National Academy of Engineering.

The Institute of Medicine was established in 1970 by the National Academy of Sciences to secure the services of eminent members of appropriate professions in the examination of policy matters pertaining to the health of the public. The Institute acts under the responsibility given to the National Academy of Sciences by its congressional charter to be an adviser to the federal government and, upon its own initiative, to identify issues of medical care, research, and education. Dr. Kenneth I. Shine is president of the Institute of Medicine.

The National Research Council was organized by the National Academy of Sciences in 1916 to associate the broad community of science and technology with the Academy's purposes of furthering knowledge and advising the federal government. Functioning in accordance with general policies determined by the Academy, the Council has become the principal operating agency of both the National Academy of Sciences and the National Academy of Engineering in providing services to the government, the public, and the scientific and engineering communities. The Council is administered jointly by both Academies and the Institute of Medicine. Dr. Bruce Alberts and Dr. William A. Wulf are chairman and vice chairman, respectively, of the National Research Council.

Support for this work was provided by Contract NASW 96013 between the National Academy of Sciences and the National Aeronautics and Space Administration.

Cover: Mars mosaic image courtesy of Alfred McEwen of the U.S. Geological Survey, Flagstaff, Arizona. Lunar crescent image courtesy of Dennis di Cicco. Cover design by Penny Margolskee.

Copies of this report are available from

Space Studies Board
National Research Council
2101 Constitution Avenue, N.W.
Washington, D.C. 20418

Copyright 1997 by the National Academy of Sciences. All rights reserved.
Printed in the United States of America

Foreword

During 1988, the National Research Council's Space Science Board reorganized itself to more effectively address NASA's advisory needs. The Board's scope was broadened: it was renamed the Space Studies Board and, among other new initiatives, the Committee on Human Exploration was created. The new committee was intended to focus on the scientific aspects of human exploration programs, rather than engineering issues. Early on, the committee recognized that an orderly review and clear statement of the role of science in human exploration should include, but distinguish between, science that must be conducted before human exploration beyond Earth's immediate environs could be practically undertaken, and science that would be enabled or facilitated by human presence on other worlds. This approach led to two reports, *Scientific Prerequisites for the Human Exploration of Space* and *Scientific Opportunities in the Human Exploration of Space*, published in 1993 and 1994, respectively. While these studies were in progress, the value of a third study that would focus on issues of science management within a human exploration program was recognized; this third topic was taken up after the *Opportunities* report was completed, and was published this year as *Science Management in the Human Exploration of Space*. These three reports are collected and reprinted in this volume in their entirety as originally published.

During the decade of existence of the Committee on Human Exploration, the prospects for human exploration have ebbed and flowed. On July 20, 1989, President George Bush announced that the United States should undertake "a sustained program of manned exploration of the solar system." Timed to commemorate the 20th anniversary of the first human landing on the Moon, this announcement formalized a deep aspiration that has suffused space enthusiasts and professionals since the very beginning of the rocket era in this century and motivated the formation of the Board's Committee on Human Exploration and its studies. Cost estimates for interplanetary travel proved very discouraging, however, and NASA's human flight capabilities were soon focused on the space station program. But the goal of flight beyond the Earth-Moon system has never entirely faded and has remained the subject of dreams and long-range studies at a low level.

The present series of reports, eight years in the making from the initial formation of the committee, seems to have been paced exactly right: the subject of human exploration of Mars is coming increasingly to the fore. Significant progress has been made in many scientific areas during this period; for example, in 1996 the Board enlarged on a key topic in the committee's first report with a detailed survey of research required in the area of biological effects of radiation.[1] Also in 1996, possible evidence for ancient Mars life was found in an Antarctic meteorite. Technology, too, has advanced enormously. As this volume goes to press, the Mars Pathfinder's Sagan Station is operating on Mars, and its tiny Sojourner rover is conducting the first mobile field geology of another planet.

Last fall, a historic partnership was formalized between NASA's human spaceflight, life science, and space science offices to collaborate in an integrated program of robotic, and ultimately human, exploration of Mars. At the same time, reinvention of NASA over the past five years has renewed commitment to developing and applying new technology and to lowering project costs. A sustained and systematic drive toward the needed science and technology may be bringing the grand challenge of human exploration of the solar system within reach.

Claude R. Canizares
Chair
Space Studies Board

Louis J. Lanzerotti
Former *Chair*
Space Studies Board

---

[1] Space Studies Board, National Research Council, *Radiation Hazards to Crews of Interplanetary Missions: Biological Issues and Research Strategies*, National Academy Press, Washington, D.C., 1996.

Contents

Scientific Prerequisites for the Human Exploration of Space

Scientific Opportunities in the Human Exploration of Space

Science Management in the Human Exploration of Space

Scientific Prerequisites
for the
Human Exploration of Space

Scientific Prerequisites
for the
Human Exploration of Space

Committee on Human Exploration

Space Studies Board

Commission on Physical Sciences, Mathematics, and Applications

National Research Council

NATIONAL ACADEMY PRESS
Washington, D.C. 1993

NOTICE: The project that is the subject of this report was approved by the Governing Board of the National Research Council, whose members are drawn from the councils of the National Academy of Sciences, the National Academy of Engineering, and the Institute of Medicine. The members of the committee responsible for the report were chosen for their special competences and with regard for appropriate balance.

This report has been reviewed by a group other than the authors according to procedures approved by a Report Review Committee consisting of members of the National Academy of Sciences, the National Academy of Engineering, and the Institute of Medicine.

The National Academy of Sciences is a private, nonprofit, self-perpetuating society of distinguished scholars engaged in scientific and engineering research, dedicated to the furtherance of science and technology and to their use for the general welfare. Upon the authority of the charter granted to it by the Congress in 1863, the Academy has a mandate that requires it to advise the federal government on scientific and technical matters. Dr. Frank Press is president of the National Academy of Sciences.

The National Academy of Engineering was established in 1964, under the charter of the National Academy of Sciences, as a parallel organization of outstanding engineers. It is autonomous in its administration and in the selection of its members, sharing with the National Academy of Sciences the responsibility for advising the federal government. The National Academy of Engineering also sponsors engineering programs aimed at meeting national needs, encourages education and research, and recognizes the superior achievements of engineers. Dr. Robert M. White is president of the National Academy of Engineering.

The Institute of Medicine was established in 1970 by the National Academy of Sciences to secure the services of eminent members of appropriate professions in the examination of policy matters pertaining to the health of the public. The Institute acts under the responsibility given to the National Academy of Sciences by its congressional charter to be an adviser to the federal government and, upon its own initiative, to identify issues of medical care, research, and education. Dr. Kenneth I. Shine is president of the Institute of Medicine.

The National Research Council was organized by the National Academy of Sciences in 1916 to associate the broad community of science and technology with the Academy's purposes of furthering knowledge and advising the federal government. Functioning in accordance with general policies determined by the Academy, the Council has become the principal operating agency of both the National Academy of Sciences and the National Academy of Engineering in providing services to the government, the public, and the scientific and engineering communities. The Council is administered jointly by both Academies and the Institute of Medicine. Dr. Frank Press and Dr. Robert M. White are chairman and vice chairman, respectively, of the National Research Council.

Support for this project was provided by Contract NASW 4627 between the National Academy of Sciences and the National Aeronautics and Space Administration.

Cover: Mars mosaic image courtesy of Alfred McEwen of the U.S. Geological Survey, Flagstaff, Arizona. Lunar crescent image courtesy of Dennis di Cicco. Cover design by Penny Margolskee.

Copies of this report are available from

Space Studies Board
National Research Council
2101 Constitution Avenue, N.W.
Washington, D.C. 20418

Printed in the United States of America

COMMITTEE ON HUMAN EXPLORATION

NOEL W. HINNERS, Martin Marietta Civil Space and Communications Company, *Chair*
RICHARD L. GARWIN, IBM T.J. Watson Research Center
LOUIS J. LANZEROTTI, AT&T Bell Laboratories
ELLIOTT C. LEVINTHAL, Stanford University
WILLIAM J. MERRELL, JR., Texas A&M University
ROBERT H. MOSER, University of New Mexico
GEORGE DRIVER NELSON, University of Washington
SALLY K. RIDE,* California Space Institute

Staff

DAVID H. SMITH, Executive Secretary
BOYCE N. AGNEW, Administrative Assistant

---

*Former committee member who participated in writing this report.

SPACE STUDIES BOARD

LOUIS J. LANZEROTTI, AT&T Bell Laboratories, *Chair*
JOSEPH A. BURNS, Cornell University
ANDREA K. DUPREE,* Harvard-Smithsonian Center for Astrophysics
JOHN A. DUTTON, Pennsylvania State University
LARRY ESPOSITO,* University of Colorado
JAMES P. FERRIS, Rensselaer Polytechnic Institute
HERBERT FRIEDMAN, Naval Research Laboratory
RICHARD L. GARWIN,* IBM T.J. Watson Research Center
RICCARDO GIACCONI, European Southern Observatory
NOEL W. HINNERS, Martin Marietta Civil Space and Communications Company
JAMES R. HOUCK,* Cornell University
DAVID A. LANDGREBE, Purdue University
ROBERT A. LAUDISE, AT&T Bell Laboratories
RICHARD S. LINDZEN, Massachusetts Institute of Technology
JOHN H. McELROY, University of Texas, Arlington
WILLIAM J. MERRELL, JR., Texas A&M University
RICHARD K. MOORE,* University of Kansas
ROBERT H. MOSER, University of New Mexico
NORMAN F. NESS, University of Delaware
MARCIA NEUGEBAUER, Jet Propulsion Laboratory
MARK SETTLE, ARCO Oil Company
WILLIAM A. SIRIGNANO, University of California, Irvine
FRED W. TUREK, Northwestern University
ARTHUR B.C. WALKER, Stanford University

MARC S. ALLEN, *Director*

---

*Term ended during 1992.

COMMISSION ON PHYSICAL SCIENCES, MATHEMATICS, AND APPLICATIONS

RICHARD N. ZARE, Stanford University, *Chair*
JOHN A. ARMSTRONG, IBM Corporation
PETER J. BICKEL, University of California, Berkeley
GEORGE F. CARRIER, Harvard University
GEORGE W. CLARK, Massachusetts Institute of Technology
MARYE ANNE FOX, University of Texas, Austin
AVNER FRIEDMAN, University of Minnesota
SUSAN L. GRAHAM, University of California, Berkeley
NEAL F. LANE, Rice University
ROBERT W. LUCKY, Bell Communications Research
CLAIRE E. MAX, Lawrence Livermore National Laboratory
CHRISTOPHER F. McKEE, University of California, Berkeley
JAMES W. MITCHELL, AT&T Bell Laboratories
RICHARD S. NICHOLSON, American Association for the Advancement of Science
ALAN SCHRIESHEIM, Argonne National Laboratory
A. RICHARD SEEBASS III, University of Colorado
KENNETH G. WILSON, Ohio State University

NORMAN METZGER, *Executive Director*

Preface

For the past 20 years, the future directions of the U.S. program of human spaceflight have been a matter of discussion, debate, and controversy within and among the government, industry, the scientific community, and the public. Many advocates of human space exploration now agree that the next steps in piloted flight after Space Station Freedom involve returning to the Moon and, eventually, voyaging to Mars. The space science community, however, is agreed that there is no a priori scientific requirement for human exploration of the Moon and Mars. This view is reflected in *Toward a New Era in Space: Realigning Policies to New Realities* (National Academy Press, Washington, D.C., 1988), a report prepared by the National Academy of Sciences and the National Academy of Engineering, which stated that "the ultimate decision to undertake further voyages of human exploration and to begin the process of expanding human activities into the solar system must be based on nontechnical factors." In that light it is proper to ask, then, what *is* a proper role for the scientific community in any program of human exploration?

Well before a human exploration program is implemented, the U.S. scientific community must involve itself by providing the scientific advice and participation necessary for enabling human exploration. Then, because virtually all mission concepts for human exploration incorporate scientific research as a major goal, it is incumbent on the research community to study how it should respond to the opportunities enabled by the existence of human exploration. The time to do that is now, for it is during the conceptualization and initial development of exploration programs that the research community has its greatest opportunity to shape the relevant political, technical, and scientific decisions. Such participation is responsive to the finding enunciated in the *Report of the Advisory Committee on the Future of the U.S. Space Program* (U.S. Government Printing Office, Washington, D.C., 1990), that science is "the fulcrum of the entire space effort."

Since its establishment in 1958, the Space Studies Board (SSB; formerly the Space Science Board) has been the principal nongovernmental advisory body on civil space research in the United States. In this capacity, the board established the Committee on Human Exploration (CHEX) to examine many of the science and science-policy matters concerned with the return of astronauts to the Moon and eventual voyages to Mars. The Board asked CHEX to consider three major questions:

1. What scientific knowledge must be obtained as a prerequisite for prolonged human space missions?
2. What scientific opportunities might derive from prolonged human space missions?
3. What basic principles should guide the management of both the prerequisite science activities necessary to enable human exploration and the scientific activities that may be carried out in conjunction with human exploration?

This report focuses on the first of these topics. Reports concerning the second and third topics are in their final stages of preparation and will be available in the near future.

The Space Studies Board and CHEX concluded that the existing research strategies of several of its discipline committees form a solid basis for determining the scientific research necessary to enable future voyages by humans to the Moon and Mars. To establish a context for its study, however, CHEX first examined the scientific aspects of various Moon/Mars mission concepts and determined the appropriate role of science in a program of human exploration. Having laid this foundation, CHEX then evaluated and integrated the *enabling requirements* for human exploration contained in the strategy documents of relevant SSB committees. (The details of the individual scientific strategies and the goals of these SSB committees are, however, not repeated in this report—they may be found in the original strategy documents listed in the bibliography.) These requirements were then classified according to their relevance to basic human survival and optimum mission performance.

Information on the conditions necessary to maintain the well-being of humans in space was provided by the Committee on Space Biology and Medicine. Requirements for data on the properties of planetary atmospheres and surfaces and exobiology, needed for basic mission operations and sci-

PREFACE   ix

ence research, were supplied by the Committee on Planetary and Lunar Exploration. *A Strategy for the Scientific Exploration of Mars* (NASA, Jet Propulsion Laboratory, Pasadena, Calif., 1991), a report written by NASA's Mars Science Working Group, was consulted for additional information on the planetological and exobiological aspects of Mars precursor science. The space radiation environment, including its characterization and predictability, is the responsibility of the Committee on Solar and Space Physics and the Committee on Solar-Terrestrial Research. Advice on some technological issues was provided by the Committee on Microgravity Research. Full membership lists for these Space Studies Board discipline committees appear in the appendix.

Contents

EXECUTIVE SUMMARY 1
  The Role of Science, 2
  Enabling Science, 3
    Critical Research Issues, 3
    Optimal Performance Issues, 4
  References, 4

1 INTRODUCTION 6
   The Human Exploration of Space, 6
   Science and the Human Exploration of Space, 8
   Enabling Science, 10
   Space Station Freedom, 13
   International Consultation and Collaboration, 13
   Notes and References, 14

2 CRITICAL RESEARCH REQUIREMENTS 15
   Radiation, 17
     Radiation Levels, 17
     Sources of Hazardous Radiation, 18
       Galactic Cosmic Radiation, 18
       Solar Energetic Particles, 20
     Relevant Measurements and Research, 22
   Bone Degeneration and Muscle Atrophy, 24

Cardiovascular and Pulmonary Function, 26
Behavior, Performance, and Human Factors, 28
   Individual Factors, 28
   Group Factors, 29
   Environmental Factors, 29
Biological Issues, 30
Notes and References, 31

3 RESEARCH FOR MISSION OPTIMIZATION   33
Sensorimotor Integration, 33
Immunology, 34
Developmental Biology, 36
Life Support Systems, 37
Micrometeoroid Flux on the Moon, 38
Surface and Subsurface Properties, 38
Potential Martian Hazards, 40
Aerobraking at Mars, 42
Microgravity Science and Technology, 43
Exobiology Issues, 43
Resource Utilization, 44
Notes and References, 45

4 CONCLUSIONS   46

BIBLIOGRAPHY   48

APPENDIX   51

Executive Summary

"To expand human presence and activity beyond Earth-orbit into the solar system"[1] was the goal established by President Ronald Reagan in 1988 for the nation's program of piloted spaceflight. This goal formed the basis for the subsequent proclamation by President George Bush on July 20, 1989—the 20th anniversary of the Apollo 11 lunar landing—in which he proposed that the nation go "back to the Moon, ... And this time, back to stay. And then—a journey into tomorrow—a manned mission to Mars."[2] The resulting long-term program to expand the human presence in the inner solar system has been called many things, including the Human Exploration Initiative, the Space Exploration Initiative (SEI), and the Moon/Mars program. The Advisory Committee on the Future of the U.S. Space Program identified these objectives as Mission from Planet Earth.[3]

It is a long way from the broad goals of human exploration to a program of implementation, with many political, technological, and scientific hurdles to be overcome. Do successive administrations and congresses, as well as the American people, have the desire to dedicate necessary national resources to support such an ambitious program? Do they have the will and patience to support a program lasting for several decades? Can humans function effectively on the Moon for long periods of time? Can they survive a lengthy mission to Mars? What will they do when they get there? These are but a few of the myriad questions to be addressed before our species can realize the ancient dream of human voyages to, and eventual settlement of, our neighboring planets.

THE ROLE OF SCIENCE

The role of science in human exploration is paramount and its challenges no less daunting than those facing the engineering community. New scientific data concerning the health and safety of astronauts are essential prerequisites for the human exploration of space. Research must be done to understand and alleviate the deleterious effects of microgravity on human physiology, the risks posed by radiation in space, and the environmental stresses humans will experience travelling to and operating on and around other planetary bodies. The U.S. scientific and engineering community is obliged to provide the best and most constructive advice to help the nation accomplish its space goals, as was stressed in a 1988 space policy report to the newly elected president by the National Academy of Sciences and the National Academy of Engineering.[4] To that end the National Research Council's Space Studies Board established the Committee on Human Exploration (CHEX) and charged it, as its first responsibility, to determine what scientific questions need to be answered before humans can undertake extended missions to the Moon and travel to Mars.

Defining these scientific prerequisites entails a degree of judgment about both our current state of knowledge of the relevant science and the potential modes of mission implementation. CHEX determined that some issues are critical to the basic survival and elementary functioning of humans in space. Other issues concern the effectiveness and efficiency of operations and their impact on overall mission success. The line between the two is sometimes fuzzy, and the committee anticipates that with time crossover will occur.

Beyond the information needed to provide for the basic health and well-being of astronauts operating in extraterrestrial environments, the expansion of human presence and activity into the solar system does not demand any a priori scientific research component. Nor is a Moon/Mars program driven by any demands for scientific discovery. The latter view is expressed in the National Academies' 1988 space policy report, which states that "the ultimate decision to undertake further voyages of human exploration and to begin the process of expanding human activities into the solar system must be based on nontechnical factors."[5] Given a nontechnical decision, what then is the proper role of science?

That *there is a role* is not open to much debate. The Paine report,[6] the Ride report,[7] the Augustine report,[8] and the report of the Synthesis Group[9] all recommend, to varying degrees, that significant scientific research be conducted in association with human exploration. In fact, "exploration" does not exist in isolation from scientific research. There are, however, two distinctly different categories of science that must be considered. There is the "enabling" science required if we are to conduct human exploration at all. Then, there is the "enabled" science made possible, or significantly enhanced, because it is carried out in conjunction with a program of human exploration. This report deals with the former topic. The latter is treated in a preliminary fashion insofar as it impacts the scientific effectiveness of Moon/Mars missions. For example, conducting certain preliminary robotic missions to the Moon and Mars can result in a more effective scientific return from eventual human exploration. This report also contains some preliminary discussion of technology requirements, aspects of international scientific cooperation, and the approach used to manage the scientific component of a program of human exploration.

ENABLING SCIENCE

In establishing the scientific prerequisites for the human exploration of space, CHEX has identified two broad categories of enabling scientific research. This classification is based on the degree of urgency with which answers are needed to particular questions before humans can safely return to the Moon or travel to Mars.

Critical Research Issues

The lack of scientific data in some areas leads to unacceptably high risks to any program of extended space exploration by humans. These critical research issues concern those areas that have the highest probability of being life threatening or seriously debilitating to astronauts and that are thus potential "showstoppers" for human exploration. The areas in which additional scientific information *must* be obtained prior to extended exploration of space by humans include the:

1. Flux of cosmic-ray particles, their energy spectra, and the extent to which their flux is modulated by the solar cycle;
2. Frequency and severity of solar flares;
3. Long- and short-term effects of ionizing radiation on human tissue;
4. Radiation environment inside proposed space vehicles;
5. Effectiveness of different types of radiation shielding and their associated penalties (e.g., spacecraft mass);
6. Detrimental effects of reduced gravity and transitions in gravitational force on all body systems (especially the cardiovascular and pulmonary systems) and on bones, muscles, and mineral metabolism, together with possible countermeasures;
7. Psychosocial aspects of long-duration confinement in microgravity with no escape possible and their effects on crew function; and
8. Biological aspects of the possible existence of martian organisms and means to prevent the forward contamination of Mars and the back contamination of Earth.

Optimal Performance Issues

The second category of research includes issues that, based on current knowledge, do not appear to pose serious detriments to the health and well-being of humans in space. They could, however, result in reduced human performance in flight or on planetary surfaces and, thus, in a less than optimal return from the mission. Some of these issues may become critical research issues relative to long-term human spaceflight and return to terrestrial gravity following extended flights, or when extraterrestrial habitation is considered. Research issues related to optimal mission performance include the:

1. Vestibular function and human sensorimotor performance;
2. Effects of the microgravity environment on human immunological functions;
3. Long-term effects of microgravity on plant growth;
4. Feasibility of closed-loop life support systems;
5. Interplanetary micrometeoroid flux and its time dependence;
6. Surface and subsurface properties of the Moon and Mars at landing sites and at the locations of possible habitats;
7. Hazards posed by martian weather and other martian geophysical phenomena;
8. Atmospheric structure of Mars relevant to implementing aerobraking techniques; and
9. Microgravity science and technology relating to long-duration spaceflight.

Two additional issues, while not directly related to human performance, are included for their potential to significantly enhance and optimize the scientific return of the mission:

10. Methods of detecting possible fossil martian organisms and the chemical precursors of life; and
11. Availability and utilization of in situ resources (e.g., ice/water and minerals) on the Moon and Mars.

REFERENCES

1. President Ronald Reagan, Presidential Directive on National Space Policy, 11 February, 1988, (Fact Sheet, page 1), The White House, Washington D.C.
2. President George Bush, Remarks by the President at 20th Anniversary of Apollo Moon Landing, 20 July, 1989, The White House, Washington D.C.
3. Advisory Committee on the Future of the U.S. Space Program, *Report of the Advisory Committee on the Future of the U.S. Space Program* (the "Augustine report"), U.S. Government Printing Office, Washington, D.C., 1990.

4. Committee on Space Policy, *Toward a New Era in Space: Realigning Policies to New Realities* (the "Stever report"), National Academy Press, Washington, D.C., 1988.
5. See Ref. 4, p. 14.
6. National Commission on Space, *Pioneering the Space Frontier,* The Report of the National Commission on Space, Bantam Books, New York, 1986.
7. Office of Exploration, *Leadership and America's Future in Space,* A Report to the Administrator by Dr. Sally K. Ride, August 1987, NASA, Washington, D.C., 1987.
8. See Ref. 3.
9. Synthesis Group, *America at the Threshold,* Report of the Synthesis Group on America's Space Exploration Initiative, U.S. Government Printing Office, Washington, D.C., 1991.

1

Introduction

THE HUMAN EXPLORATION OF SPACE

On July 20, 1989, President George Bush set an ambitious vision before the American people: to go "back to the Moon, ... And this time, back to stay. And then—a journey into tomorrow—a manned mission to Mars."[1] This proposal to expand human presence in the solar system has been given a number of different names, including the Human Exploration Initiative, the Moon/Mars program, Mission from Planet Earth, and, most recently, the Space Exploration Initiative (SEI). In this report, the term "Moon/Mars program" is used to refer generically to any future program directed toward the human exploration of the Moon and Mars.

In the last decade, many committees, commissions, and studies have assessed the future of the U.S. space program and have come to broadly similar conclusions regarding the future of human spaceflight. The most recent major assessment, performed by the Stafford Commission (or Synthesis Group) in a report[2] submitted to Vice President J. Danforth Quayle on May 3, 1991, set forth six defining themes to guide human exploration:

1. Increase our knowledge of the solar system and the universe;
2. Rejuvenate interest in science and engineering;
3. Refocus the U.S. position in world leadership away from the military to the economic and scientific spheres;
4. Develop technology that has terrestrial application;

5. Facilitate further space exploration and commercialization; and
6. Boost the U.S. economy.

The fundamental premise of a Moon/Mars program, given the overarching goal of human presence and activity beyond Earth, is directly articulated by the first theme, an increase in knowledge of the universe. Thus "the Space Exploration Initiative is an integrated program of missions by humans and robots to explore, to understand and to gain knowledge of the universe and our place in it."[3]

As its name suggests, the Synthesis Group's report was the distillation of a nationwide outreach campaign to ascertain the nation's space exploration aspirations. The group devised four broad concepts, or architectures, each embodying an alternative goal. The first emphasizes an accelerated human mission to Mars, with an intermediate return to the Moon. The second concentrates on scientific research on the Moon and Mars. The third provides for long-term habitation on the Moon, accompanied by a Mars exploration phase. The final architecture envisages the utilization of in situ lunar and martian resources to expand human capabilities in the inner solar system.

The report of the Synthesis Group proposed a strategic approach with its use of "waypoints." Each waypoint describes a level of capability that is, in itself, a significant achievement. At each waypoint the accumulation of infrastructure, technology, and knowledge would allow selection of both the emphasis and detailed implementation needed to achieve the next waypoint. The architecture is thus an assemblage of successive waypoints.

While not intended as detailed blueprints for the execution of a program of human exploration, the architectures characterize broad alternative goals for a Moon/Mars program. Science plays a major, albeit different, role in each concept. However, certain recurring scientific elements are found in all four architectures and, incidentally, in previous studies of the human exploration of space. These common themes include the following:

- The principal barriers to human exploration, particularly of Mars, are uncertainties in medical science. These uncertainties include, in particular, the physiological and psychological burdens placed on the crews and the acceptable level of risk that can be assumed;
- A mix of robotic and human exploration missions. The former (precursors) may provide information necessary for the planning and successful execution of the latter or may undertake purely scientific tasks (although the report of the Synthesis Group did not emphasize their scientific potential);
- Initial human activities on the Moon. Some are specifically preparatory for Mars missions. Others deal with study or use of the Moon for science;

- The prime objectives of the Mars missions are exploration and science; and
- Significant technical advances are required if humans are to return to the Moon and travel on to Mars. These are primarily engineering developments of existing or understood technologies rather than the development of totally new scientific or technological approaches.

The Space Studies Board's Committee on Human Exploration (CHEX) presumes that, eventually, one of these architectures, or perhaps even a new theme, could be selected to provide a focus for Moon/Mars exploration. Once this is done, the subordinate objectives can be deduced and mission planning begun.

Regardless of which specific architecture is ultimately selected, human exploration of the Moon and Mars will be a long-term program of progressively more complex and demanding missions. These will challenge the nation's technical capabilities, management skills, and, perhaps, financial resources.

SCIENCE AND THE HUMAN EXPLORATION OF SPACE

Ever since the successes of the Apollo program 20 years ago, the future directions of the U.S. program of human spaceflight have been a matter of discussion, debate, and controversy within the government and the scientific community and among the public. A report on space policy by the National Academy of Sciences and the National Academy of Engineering stated that "the ultimate decision to undertake further voyages of human exploration and to begin the process of expanding human activities into the solar system must be based on nontechnical factors."[4] Nevertheless, the U.S. research community is obliged to provide the best and most constructive scientific advice it can to shape the political and technical decisions regarding piloted flight. This role is consistent with the recommendation of the Augustine Committee that science is "the fulcrum of the entire space effort."[5]

Part of the task facing the scientific community is determining what knowledge is prerequisite for prolonged human space missions. However, these prerequisites depend on the goals of such missions. If the goal of future space missions were solely to satisfy the "human imperative" to explore or to enhance national prestige or other nontechnical and nonscientific objectives, there would be a limited set of requirements. There would, for example, be relatively little need for precursor robotic missions to characterize the martian surface, because sufficient data are at hand from the Viking mission to allow selection of a safe landing site. But because the goals of most Moon/Mars concepts to date do include the expansion of knowledge and other objectives such as long-term habitation and utilization of in situ resources, the set of prerequisites is larger. For example, a martian landing site must not only be safe but must also be desirable from a scientific perspective. This creates a need for precursor robotic missions and provides linkages between the scientific knowledge that is prerequisite for human exploration and the scientific opportunities deriving from such a program.

The relative role of humans and robotic probes in space exploration has long been a contentious issue. If the acquisition of knowledge were the only goal, then the criteria for selecting between humans and robots would be clear: select the most cost-effective method of obtaining the desired results. The Augustine report recognized the important role humans can play in exploration. However, it went on to say that "in hindsight . . . it was . . . inappropriate in the case of the *Challenger* to risk the lives of seven astronauts and nearly one fourth of NASA's launch assets to place in orbit a communications satellite."[6] A rational approach is to use robots until we can define objectives for which humans are essential. We could also conduct experiments to determine the contribution to field exploration that is gained by having humans in situ. No compelling case has yet been made that human exploration is necessary to accomplish the goals of lunar and martian science or, for that matter, any other goal except the "human imperative" to explore. The report of the Synthesis Group gives five visions other than science. However laudable these other visions are, there has been no cost-benefit analysis to show that human exploration is the best way of achieving them.

The tension between the science and nonscience goals suggests the following criteria for selection between human and robotic options. Robotic probes should be used to provide enough information to:

1. Optimize the sites chosen for human exploration. Mars especially, but also the Moon, presents varied environments, and the number of sites astronauts can visit will be limited, as will be the range of their traverses at each site; and

2. Define a set of scientifically important tasks that can be *well* performed by humans in situ.

The first criterion should not be interpreted to mean that there is currently a scientific justification for human exploration. Nor does the second demand (at least initially) that scientific tasks would be best and most cost-effectively performed by humans. It is possible that future experiments and flight experiences will show that some tasks are better, and perhaps more cost-effectively, performed by humans, given the state of the art of robotic technology. If this should turn out to be the case, a scientific justification for human exploration might evolve.

The inclusion of science goals in a Moon/Mars program raises two serious concerns for the scientific community. The first is that human exploration may displace other programs and initiatives that have a higher scientific significance or priority. The second concern is that the scientific objectives be of high quality and be competitive with other scientific opportunities. Toward this end, the scientific component of human exploration should be managed so that:

1. The stated scientific objectives of the human exploration program are achievable with a high probability of success;
2. The architecture is flexible and able to respond to new scientific discoveries and, thus, to ensure that the scientific benefits of the program are maximized;
3. Scientific advice is included in day-to-day decisions on the strategy and implementation necessary to execute the programs; and
4. All goals (e.g., scientific research, human presence, utilization of resources) of a Moon/Mars program are clearly stated and represented in project management in such a manner that open and effective decision making can be accomplished.

Management issues will be dealt with in depth in the third CHEX report; they are mentioned here to emphasize the necessity to deal with the approach to science management ab initio.

ENABLING SCIENCE

A Moon/Mars program requires the acquisition of scientific data either prior to, or in conjunction with, actual piloted flight and planetary surface activity. Establishing the requirements for such data is, to a major extent, a task for the scientific community. This entails both a responsibility and an opportunity. The responsibility is to state clearly what scientific data are essential to enable a Moon/Mars program and to propose programs and mechanisms to acquire, analyze, and interpret data, and to assure the overall quality of the scientific research. An opportunity arises because some enabling data will have a value over and above that immediately required by a program of human exploration. Such information might, however, be accorded a different priority in the absence of a program of human exploration.

Developing the full set of requirements for enabling data is an iterative process that will depend eventually on the specific architecture selected. If, for example, establishing astronomical observatories on the Moon becomes a goal, particular information on the lunar environment that might otherwise not be needed will become essential. Similarly, if long-term habitation becomes a goal of lunar or martian exploration, then the search for in situ resources such as water becomes a high priority simply because of the major impact that easily recoverable resources would have on the entire program.

Conversely, consideration of a set of enabling requirements derived from a particular architecture, and the ability to satisfy those requirements, could produce changes in the architecture. For example, if it turns out to be impossible to devise countermeasures to the deleterious effects of long-term exposure to microgravity, then the development of a vehicle that incorporates artificial gravity or the development of advanced propulsion systems with decreased transit times may be the only practical options.

Scientific information is clearly needed to assure the safety of humans and the effectiveness of human and machine operations. Although the Apollo missions have proven that humans can undertake brief expeditions to the Moon, the prospect of long-term or permanent habitation raises serious safety issues, particularly where current knowledge is only rudimentary. Apollo data provide some clues as to areas in which our ignorance harbors the greatest potential dangers. These areas include the long-term and short-term prediction of solar flares,[7] the character of the interplanetary meteoroid flux, the detailed nature of the lunar subsurface, and the possible detrimental effects of long-term interaction with the ubiquitous lunar dust.

Some of the basic knowledge about the atmosphere and surface of Mars required for human exploration is already in hand. The United States successfully operated two robot landers for more than one martian year. Yet, despite the wealth of data gathered by the Viking probes, extensive human activities on Mars will require the acquisition of significant amounts of new information. The variability of the martian atmosphere, the planet's surface and subsurface characteristics, and the risk of volcanic activity must be studied. The existence or abundance of significant, life-critical resources needs to be determined. Attention must be given to avoiding the transport of microorganisms from Earth and vice versa. The identification of likely abodes of any past life will follow from a better understanding of the martian environment and its history.

Precursor robotic missions (including sample return missions) can permit analyses that would greatly improve the selection of landing and exploration sites that could, in turn, enhance the science to be accomplished by human exploration. A Mars sample return mission may be desirable to settle questions of forward contamination and back contamination. Indeed, the Space Studies Board has recommended that "the next major phase of Mars exploration for the United States involve detailed in situ investigations of the surface of Mars and the return to Earth for laboratory analysis of selected martian surface samples."[8]

In examining the enabling science for the human exploration of space, CHEX identified two categories of research topics, each with differing degrees of urgency. *Critical research issues* are those related to conditions known to be life-threatening or seriously debilitating: they are the potential "showstoppers" of human exploration. The other category, *research for mission optimization*, includes issues that, based on current knowledge, do not appear to represent immediate threats to the health and well-being of humans in space. They could, however, result in reduced astronaut performance in flight or on the surface of the Moon or Mars, leading to a suboptimal mission. They could also impact the health of astronauts long after a mission is completed. In addition, it must be recognized that our current state of ignorance about prolonged human spaceflight leaves open the possibility of phenomena that cannot be anticipated.

CHEX emphasizes that, as new information is acquired, some optimal performance issues could become critical to ensuring the well-being of astronauts. If, for example, it is necessary to minimize payload mass, development of a partially closed, if not fully closed, life support system could become mandatory for missions to Mars.

The exploration of Mars by humans will be one of the most complex, challenging, and expensive technical endeavors ever attempted. These missions will, however, be carried out by even more complex entities—humans. It is therefore vital that as much effort be put into understanding the effects of the space environment on humans as has been put into understanding the mechanisms of getting a spacecraft to Mars and back.

It is widely assumed that since a small number of astronauts have survived and operated for as long as a year in space, there are no major physiological problems that would prohibit long-term human exploration. This assumption is unwarranted. An assessment of current research in space biology and medicine shows that the major problems posed by prolonged exposure to microgravity remain no nearer solution in 1993 than they were in 1961, the year of the first human spaceflight. For reasons outlined in earlier reports,[9] space biology and medicine are in the very earliest stage of development as rigorous scientific disciplines. These fields must mature if any attempt is made to send humans on extended missions to Mars.

The danger posed by biomedical uncertainties is related to another important matter, not often publicly stated—the role of courageous individuals. Humans who venture into space must accept a degree of personal risk. But, as the *Challenger* accident made clear, the public will not accept losses that can be anticipated and avoided. A sustained program of human exploration must adopt the prudent strategy of reducing to an acceptable minimum both the immediate and long-term risks astronauts will face. Thus, the potential hazards of exposure to radiation and microgravity must be addressed within the context of a comprehensive program of health and safety. To do otherwise imposes unacceptable risks on the entire human exploration enterprise.

SPACE STATION FREEDOM

What role does Space Station Freedom play in the future human exploration of space? The Augustine report recommended that the primary objective of a space station should be life sciences research.[10] The Space Studies Board strongly affirms the position that a suitably equipped space-based laboratory is required to study the physiological consequences of long-term spaceflight.[11] The 1987 report of the Space Studies Board's Committee on Space Biology and Medicine laid out the critical requirements for such a space station.[12] They include:

1. A dedicated life sciences laboratory with adequate crew to conduct research;
2. A variable-speed centrifuge of the largest possible dimensions;
3. Sufficient numbers of experimental subjects (humans, plants, and animals) to address the stated scientific goals; and
4. Sufficient laboratory resources, including power, equipment, space, computational facilities, and atmosphere, to support the above research requirements.

NASA's current plans for Space Station Freedom are the subject of much controversy because of the project's escalating cost, lengthening construction schedule, and declining capabilities. On several occasions, the Space Studies Board has expressed concern that the current, descoped design of Space Station Freedom does not meet all the basic research requirements outlined above[13] and therefore will *not* fulfill its role as the first and necessary step in the human exploration of space. This is especially true if we are to use Space Station Freedom to perform the necessarily long program of enabling biomedical research and still meet the oft-stated goal of landing humans on Mars by 2019. The prudent strategy is, as the Augustine report recommended, to be flexible and not set a rigid schedule for the exploration of Mars by humans. However, the difficulties currently being experienced by the space-station project do not negate the essential need for such a facility to perform the enabling research on human adaptation to the microgravity environment necessary for a Moon/Mars program.

INTERNATIONAL CONSULTATION AND COLLABORATION

The magnitude and comprehensive nature of a Moon/Mars project will present unprecedented opportunities for cooperation with other nations. Just as other countries will play important roles in building the spacecraft and systems to support human exploration, so too will they be intimately involved in both the scientific research necessary to enable human exploration of the Moon and Mars, and in the enabled science opportunities arising from such explorations.

To a great degree, space science is already broadly international. A multitude of mechanisms exist for involving the most creative minds around the world in space science, from canvasing the international community to determine scientific objectives to inviting participation in specific missions. Just as the space hardware programs of other countries have matured, so also have their space science capabilities; thus they will expect to be treated as equal, not junior, partners in the human exploration enterprise. CHEX believes, therefore, that a consensus of the international space-research community on the scientific goals and objectives of a Moon/Mars program, and on a strategy for their implementation, is essential to the development of any framework for cooperation in the overall human exploration program.

NOTES AND REFERENCES

1. President George Bush, Remarks by the President at 20th Anniversary of Apollo Moon Landing, 20 July, 1989, The White House, Washington D.C.
2. Synthesis Group, *America at the Threshold,* Report of the Synthesis Group on America's Space Exploration Initiative, U.S. Government Printing Office, Washington, D.C., 1991.
3. See Ref. 2, p. 2.
4. Committee on Space Policy, *Toward a New Era in Space: Realigning Policies to New Realities* (the "Stever report"), National Academy Press, Washington, D.C., 1988, p. 14.
5. Advisory Committee on the Future of the U.S. Space Program, *Report of the Advisory Committee on the Future of the U.S. Space Program* (the "Augustine report"), U.S. Government Printing Office, Washington, D.C., 1990, p. 5.
6. See Ref. 5, p. 3.
7. It is worth noting that the crews of Apollos 16 and 17 were very lucky in that their flights bracketed the large flare of August 4, 1972. If the mission timings had not been so fortuitous, the astronauts could have suffered potentially fatal exposure to radiation.
8. Space Studies Board, *International Cooperation for Mars Exploration and Sample Return,* National Academy Press, Washington, D.C., 1990, p. 1, p. 3, and p. 25. See also, Space Studies Board, *1990 Update to Strategy for the Exploration of the Inner Planets*, National Academy Press, Washington, D.C., 1990, p. 40, and Space Science Board, *Strategy for Exploration of the Inner Planets: 1977–1987,* National Academy of Sciences, Washington, D.C., 1978.
9. See, for example, Space Science Board, *A Strategy for Space Biology and Medical Sciences for the 1980s and 1990s,* National Academy Press, Washington, D.C., 1987.
10. See Ref. 5, p. 29 and p. 47.
11. Space Studies Board, *Assessment of Programs in Space Biology and Medicine 1991,* National Academy Press, Washington, D.C., 1991.
12. See Ref. 9, pp. 13-16.
13. See Space Studies Board, *Space Studies Board Position Paper on Proposed Redesign of Space Station Freedom,* March 1991, and *Space Studies Board Assessment of the Space Station Freedom Program,* March 1992.

2

Critical Research Requirements

The cardinal consideration in any discussion of prolonged human exploration is the safety and well-being of the crew. This led CHEX to define a set of critical research requirements related to conditions known to be life threatening or seriously debilitating: they are the potential "showstoppers" of human exploration. All previous experience from Mercury to the Space Shuttle and from Vostok to Mir is helpful in indicating possible problems. This experience is, however, insufficient to provide all the answers about the long-term effects of spaceflight on humans, since that experience is limited to less than three months for U.S. astronauts (almost 20 years ago) and just over one year for a small number of cosmonauts. In addition to the limited time, many of the effects were inadequately studied from a research protocol point of view.

In contemplating round-trip voyages to Mars of two years or more, we enter a new arena of human experience. Factors such as radiation, the effects of prolonged exposure to microgravity on physiologic functions, the psychosocial phenomenon of sequestration of a small crew in a confined area, with a closed environmental system and without any prospect of escape in the event of catastrophe, are all without precedent.[1] Ground-based research characterizing the effects of psychosocial and radiation phenomena should be continued and enhanced.

Space biology and medicine are in such a primitive state of development that knowledgeable researchers cannot state with any degree of assurance that human crews will be able to operate their spacecraft or function usefully on Mars after their voyage. Even if nuclear- or solar-thermal (or nuclear- or solar-electric) propulsion systems can be realized, trip time will still be nearly six months each way. Even this is well beyond U.S. experience, and the former-Soviet Union's program offers very limited solid biomedical data for missions of this duration.

Once astronauts reach their destinations, they may face additional problems. We have no information at all about the physiological effects of long-duration (more than one year in some scenarios) exposure to the fractional-g lunar or martian environments. One recent report asserts that "it is expected that while crews are on the martian surface, the three-eighths Earth's gravity will help maintain their physiological health."[2] There is absolutely no scientific evidence to support this expectation.

Some space planners are optimistic that essential information can be obtained and necessary measures taken to ensure reasonable safety for crew members. In the view of CHEX this is far from a certainty. Thus life-sciences research must be the dominant factor in any consideration of prolonged human spacefaring. All other aspects of a Moon/Mars program fade into secondary importance until the relevant life-sciences research has been conducted and preventive or ameliorative measures investigated. It is critical that planners recognize that current knowledge about human performance in space is predicated on relatively short-term experiences. CHEX predicts that human problems that we cannot anticipate today will be discovered during long-term missions.

It has been suggested that some of the enabling biomedical data can be gained in operations conducted on the Moon.[3] Such operations will not, however, be sufficient to yield the biological and physiological information required for a comprehensive understanding of the effects of microgravity. There can be no assurance that countermeasures derived in an ad hoc manner will be effective for all crew members in all situations.

CHEX recommends that those implementing a Moon/Mars program commit to and lead a comprehensive program of basic and applied life-sciences research on the effects on human physiology of the microgravity, reduced-gravity, and space-radiation environment prior to finalizing spacecraft designs or undertaking long-duration flights. For this purpose, a long-term research program in adaptation to microgravity and reduced gravity, properly conducted in a suitably equipped space station in low Earth orbit, will be required. Such a research program may require 5 to 10 years because of the necessarily long duration of individual experimental protocols.

RADIATION

Bombardment by energetic particles is a major hazard facing space travellers.[4] Indeed, NASA has recognized that the cumulative radiation dose "will probably be the ultimate limiting factor for human exploration."[5]

Humans conducting extended space voyages face two different radiation hazards: a protracted exposure to galactic cosmic rays at a low dose rate and some probability of exposure to considerably higher doses of solar energetic particles. Depending on the total exposure suffered, these twin effects will increase the probability of stochastic effects (such as cancer and genetic damage) and may also increase the incidence of deterministic effects (physical damage to tissues). The effects of acute irradiation during solar particle events are of particular concern. The high-dose-rate exposures they could inflict on astronauts could cause acute damage to the skin, gut, bone marrow, and germinative tissues and, at a later date, cause cataracts. Estimating the probability of very large solar flares and predicting the resultant exposure of astronauts to radiation are among the principal concerns that need to be addressed before we can safely design new space vehicles and plan voyages of human exploration.

Radiation Levels

The health hazard posed by energetic particles depends, in part, on the energy deposited as the particles pass through tissue or come to rest in vital organs. This is traditionally characterized by the "dose equivalent," which reflects the biological effect of exposure to radiation. The dose equivalent is equal to the absorbed dose multiplied by the "quality factor" (Q), which varies from ~1 for minimally ionizing particles such as gamma rays to ~20 for neutrons and heavy ions such as iron nuclei.

The International Commission on Radiological Protection has recently recommended that the term "quality factor" be replaced by "radiation weighting factor" (W_R). The values of W_R for specific types and energies of radiation have been selected to be representative of the relative biological effectiveness (RBE) of radiation in inducing stochastic effects at low dose.[6] There are, however, no recommendations for values of W_R for causing either early or late deterministic effects such as acute tissue damage and cataracts, respectively. However, the RBE for cell killing by radiation with high linear-energy-transfer rates (e.g., heavy ions and neutrons) is considerably lower (by factors of about two to five) than that for the induction of cancer.

NASA currently has no limits for exposure to radiation during deep-space missions conducted beyond the protective shield of the geomagnetic field because little is known about the physiological effects of the heavy ions found in cosmic rays. In terms of the traditional dose-equivalent formulation, NASA's current limits for exposure of astronauts in low Earth orbit are 0.25 sievert (Sv)[7] per month, 0.5 Sv per year, and 1 to 4 Sv for a lifetime exposure (depending on age and sex). (For comparison, the typical dose used to sterilize food and drugs is 20,000 Sv.) NASA's current limits correspond to a 3% excess risk of eventual death due to cancer and are about 10 times that allowed for terrestrial radiation workers and about 100 times that allowed for the general population.

Sources of Hazardous Radiation

As mentioned above, two types of radiation are hazardous to astronauts—galactic cosmic rays and solar energetic particles. The risk posed by galactic cosmic rays is principally due to protons (with a broad range of energies) and heavy ions (in particular, energetic iron nuclei). The principal danger from solar energetic particles is posed by sporadic, large fluxes of energetic protons.

Galactic Cosmic Radiation

Galactic cosmic rays consist of ions of all atomic numbers from 1 to 92, with energies ranging up to 10^{20} electron volts (eV). Those combining high (H) atomic number (Z) and high energy (E) are collectively called HZE particles. Of these, the iron-group ions are the most hazardous because they combine relatively high abundance, a high rate of energy deposition (proportional to the square of their electric charge), and a high Q-factor. To a lesser extent, ions with atomic numbers between those of oxygen and silicon are also important.

Many questions concerning HZE particles are unanswered. How effective are they, for example, in inducing cancers? Can the late deterministic effects of HZE particles be predicted from our present understanding of the long-term effects of radiations with low linear-energy-transfer rates such as x rays and gamma rays?

Two other areas where more data are needed are of particular relevance to human exploration. The first is the 10 to 30% range of uncertainty in the measured fluxes of heavy ions in the critical energy range from 50 to 5000 MeV per nucleon (which includes more than 90% of the cosmic-ray flux). Second, the Sun's 11-year activity cycle modulates the cosmic-ray flux such that the flux at energies below 5000 MeV per nucleon is greater during years of solar-activity minimum than during solar maximum. As alluded to previously, a better understanding of the biological effects, both acute and long term, of energetic radiation must also be achieved.

The materials that form the spacecraft or the layers of a spacesuit shield astronauts from radiation to some extent. In addition, the human body

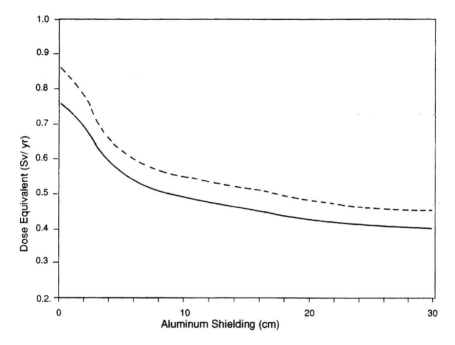

FIGURE 1 Estimates (solid curve) of the radiation dose equivalent received from galactic cosmic rays at a depth of 5 cm in body tissue (representative of, for example, bone marrow) versus aluminum shielding thickness during the 1977 solar-activity minimum. The dashed curve is an upper bound on the dose equivalent at the 90% confidence level. From Adams et al., 1991 (see reference 8).

provides ~5 cm of additional shielding for *some* critical organs, which is equivalent to about 4 cm (or 10 gm/cm^2) of aluminum. Figure 1 illustrates the estimated dose equivalent at 5-cm tissue depth for aluminum shielding of different thicknesses for galactic cosmic rays during the cosmic-ray maximum (solar-activity minimum) in 1977.[8] As can be seen, only the first 5 cm of shielding is very effective; disproportionately thicker shields are required for greater protection. For comparison, one third of the solid angle inside the space shuttle has a shielding of less than 8 cm of aluminum, while 11% of the solid angle has a shielding equivalent to less than 0.8 cm of aluminum.[9] In addition to attenuating the flux, the thickness and type of shielding determine how cosmic rays fragment into secondary particles. The nature and abundance of these secondaries, which account for the flattening of the dose-versus-shielding curve, are a major determinant of the radiation dose astronauts will receive.

The great penetrating power of cosmic rays combined with their high RBE suggests it may be impractical to shield against them in deep space.

Therefore, if background cosmic rays were the only radiation hazard, the safest time for a mission to Mars might be when the Sun's activity is near maximum and the flux of galactic cosmic rays might be 10 to 30% lower due to the modulation effect. Unfortunately this time corresponds to the period of the highest probability of solar-flare occurrence. Thus, a voyage to Mars during solar maximum should be conducted only if timely forecasts of solar energetic particle events will exist to allow adequate defensive measures to be taken. Before any final conclusions on mission timing are drawn, the probability of solar-flare occurrence must be considered along with the uncertainties in cosmic-ray fluxes, their modulation, attenuation, and fragmentation in shielding, and biological effects.

Solar Energetic Particles

The intensity, spectra, and composition of energetic particles from solar flares are much more variable than those of galactic cosmic rays. The flare-produced energetic-particle population can also be dramatically enhanced by strong shocks in the solar wind associated with coronal mass ejection. An unprotected astronaut caught in a very large flare event could be exposed to a very high or even a lethal dose in a few hours to a day. The most dangerous events are those that include solar protons with energies above a few tens of MeV. The alpha particles, electrons, and heavier nuclei accompanying the protons pose comparatively slight additional hazards.

Shielding can provide some degree of protection against solar energetic particles. Figure 2 shows the effectiveness of aluminum shielding for the large flare of August 1972 and a hypothetical "worst case" combining the very-high-energy particles observed in the February 1956 event with the very high flux levels attained in the August 1972 event.[10] As can be seen, a worst-case event would place astronauts at considerable risk because of their prolonged exposure to energetic protons at relatively high dose rates even if they were shielded by 16 cm of aluminum. It must be noted that detailed measurements of solar flares have been available for only a few decades, and so events with characteristics even more extreme than this "worst case" cannot be excluded with any confidence.

A lunar or martian base could be partially buried so that its inhabitants would be protected from radiation when inside. They would, however, still be at risk in transit between Earth, the Moon, and Mars and when on the lunar and martian surfaces. Thus space travellers will likely need some type of early warning system to alert them to dangerous solar events. In addition, mission rules would need to take into account the time needed to seek shelter.

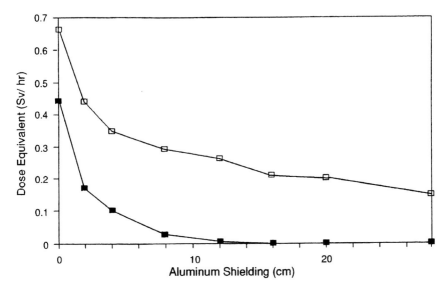

FIGURE 2 The radiation dose equivalent received at a depth of 5 cm in body tissue (representative of, for example, bone marrow) versus aluminum shielding thickness for the August 1972 solar flare (solid squares) and a composite, worst-case solar energetic particle event (open squares). Reprinted with permission from J.R. Letaw, R. Silberberg, and C.H. Tsao, "Galactic Cosmic Radiation Doses to Astronauts Outside the Magnetosphere," in *Terrestrial Space Radiation and Its Biological Effects*, P.D. McCormack, C.E. Swenberg, and H. Bucker (eds.), Plenum Press, New York, 1988. Copyright 1988 by Plenum Publishing Corp.

In addition to hazardous energetic particles, solar flares produce energetic neutrons and enhanced electromagnetic emissions at all wavelengths. Although the increased radio, optical, ultraviolet, and x rays do not constitute a hazard, they do signal the onset of proton acceleration in the Sun. This electromagnetic radiation travels at the speed of light and takes only eight minutes to reach the Earth-Moon system in contrast to energetic solar-flare protons, which may take from 15 minutes to 60 hours to travel the same distance.[11] Thus, a flare-radiation detection system could give adequate warning for crews working near a lunar base. For astronauts engaged in surface traverses on the Moon or Mars, emergency procedures must be developed to provide temporary shielding rapidly. Orbital transfer vehicles will need storm shelters where crew members can take refuge during an event. The need for emergency procedures will tend to be minimized if dangerous flare conditions can eventually be predicted a day or more in advance.

Relevant Measurements and Research

There are several possible approaches for making significant progress in reducing some of the current uncertainties in the flux of heavy ions in galactic cosmic rays. These include the following:

• The fluxes of cosmic-ray nuclei (especially oxygen through iron) should be measured throughout the 22-year magnetic solar cycle using a new generation of instruments with large geometric factors, such as NASA's planned Advanced Composition Explorer;
• Measurements of the intensities of the electron and positron components of galactic cosmic rays over most of a 22-year cycle would separate charge-sign-dependent effects from other cosmic-ray propagation effects, thereby leading to better understanding of the modulation process;
• Measurement of the galactic cosmic-ray intensities beyond the boundary of the heliosphere would establish an upper limit to the radiation intensity independent of its modulation by the solar wind and magnetic field. Continued tracking of the Voyager spacecraft is clearly cost-effective in this respect; and
• Theoretical studies of the solar- and plasma-physical processes that modulate the intensity of galactic cosmic rays are required to better understand and predict their variability.

Improved measurements of cross-sections and better modeling of heavy-ion interactions, particularly for the yield and spectra of neutrons and other secondary particles generated in the shielding material, are also required. NASA currently helps support the Bevalac heavy-ion accelerator and some cross-section studies. However, the Bevalac has been threatened with closure, thus endangering some of the enabling research on both cross-section measurements and the long-term biological effects of ionizing radiation.[12]

Research conducted during the International Geophysical Year in the late 1950s helped lay the groundwork for the basic theoretical understanding of the triggering of solar flares: fast magnetic reconnection in a magnetically dominated plasma. Since then, progress in understanding the details of the solar-flare mechanism has been slow. Moreover, in the absence of human spaceflights beyond low Earth orbit, flare prediction has not been the focus of solar-flare researchers for the last 15 years. There is, however, reason to believe that significant progress can be made if the objectives are compelling.

Two types of research programs should be considered: first, those that help us understand the process of particle acceleration and release and that might eventually lead to improved forecasting of energetic-particle events, and second, those that provide warning that a potentially dangerous event has occurred.

In the first category, the following programs would lead to significant progress in understanding flares:

- A meter-class space telescope to observe the Sun continuously with 100-km resolution. This facility should advance our fundamental understanding of flare-production mechanisms by spotting such precursor events as the emergence of magnetic flux through the photosphere and the buildup of magnetic shear;
- A global network of some 6 to 10 small Earth-based solar telescopes to measure magnetic fields and optical radiation over the full solar disk with approximately 700-km resolution. By monitoring active regions and logging flare precursors, these instruments should lead to better flare forecasting on time scales of hours to days;
- An x-ray and gamma-ray imaging telescope in space to provide information on the acceleration and propagation of energetic electrons and ions in the flare plasmas, and hence on the nature of the flare process. When coupled with direct and proxy measurements of the evolution of the magnetic-field structure in the flaring regions, this could substantially increase our ability to predict the acceleration and release of energetic flare particles; and
- Theoretical studies and computer simulations of flare-related magnetohydrodynamic processes to interpret the required measurements and direct future observations.

Whether or not we are ever able to forecast flares with high confidence, the following space-based measurements could be used as part of an advance-warning system for energetic particles once a flare has occurred.

1. A solar-observing spacecraft stationed 1 astronomical unit from the Sun in solar orbit 60 to 90 degrees ahead of Earth. Its payload would consist of an extreme-ultraviolet/x-ray telescope, a white-light coronagraph, and a small telescope designed to detect the onset of flares.
2. A network of satellites spaced at 90-degree intervals in a solar orbit with a radius of 0.3 to 0.5 astronomical unit. These satellites would carry energetic-particle detectors to provide reliable early warnings of energetic flare particles.

A solar-observing spacecraft is an important component of a short-term (a few minutes to a few hours) warning system because it would allow modeling and predictions of the paths taken by energetic particles as they are channeled from flare sites into interplanetary space.

The coronagraph would allow coronal mass ejections (CMEs) to be observed and their initial speeds to be determined. Such observations provide 1- to 3-day advance warning of the arrival of the CME-driven shocks that can dramatically enhance the population of flare-produced energetic particles. A single spacecraft is sufficient to cover the Earth-Moon system, but a network of three or four spacecraft (with 90- to 120-degree spacing) is required to cover Mars exploration, because Earth and Mars have different orbital periods and solar longitudes.

BONE DEGENERATION AND MUSCLE ATROPHY

Microgravity has major, potentially dangerous effects on human physiology. Extensive research is required to understand the responses of humans to microgravity and to assess their implications for long-duration spaceflight. Because a small number of astronauts and cosmonauts have survived long-duration missions in low Earth orbit, there is a false perception that there is no need to be concerned about health-related issues when contemplating interplanetary voyages. According to the Committee on Space Biology and Medicine, "Based on what we know today, this assumption of continued success cannot be rigorously defended."[13] The committee continued, "If this country is committed to a future of humans in space, particularly for long periods of time, it is essential that the vast number of uncertainties about the effects of microgravity on humans and other living organisms be recognized and vigorously addressed. Not to do so would be imprudent at best—quite possibly, irresponsible."[14]

The bone degradation (osteopenia) and muscle atrophy that occur in a microgravity environment are severe hurdles to an extended human presence in space.[15] The primary risk is to the functioning of the musculoskeletal system upon reexposure to planetary gravity. At present, our understanding of the causes of space-induced osteopenia and muscle atrophy is inadequate to devise effective countermeasures to be taken on long-duration space missions. Also lacking are data on the temporal sequence of bone remodeling and muscle atrophy in prolonged exposure to microgravity and the ways in which these processes may depend on other risk factors such as age, gender, race, or nutrition. Without such data, we cannot be confident that a prolonged microgravity mission such as a Mars flight would not lead to irreparable musculoskeletal damage. Such damage could both impair the effectiveness of crew members during their stay on Mars and pose serious problems upon their return to Earth. There is also the possibility that some bone demineralization will occur during prolonged flight in spite of countermeasures. If so, astronauts en route to Mars might be at risk for bone fracture with mild trauma and for the formation of kidney stones.

There is great depth and breadth to current research on osteopenia, muscle atrophy, and their underlying causes, thanks to sponsorship by the National Institutes of Health. These studies have concentrated on the problems of bone metabolism in relation to aging, menopause, endocrine disorders, poor nutrition, immobilization, and extended bed rest. A major effort is now needed to develop parallel studies to acquire basic knowledge about these problems as they occur in microgravity and to begin devising appropriate countermeasures. A critical factor in such studies must be the use of appropriate animal models and the development of computational and experimental methodologies to test and validate mechanisms of bone remodeling and muscle conditioning. In addition, the development of suitable in vitro systems using bone and muscle tissue cultures should be undertaken.

One approach to counteracting the physiological effects of microgravity is to subject organisms in space to artificial gravity. Although such an environment could correct bone degeneration, muscle atrophy, and other changes due to microgravity, it could also exacerbate other effects not now perceived to be major problems. Head movements made in a spinning environment or Coriolis effects can lead to disturbing vestibular sensations and motion sickness. Changes in gravity experienced when moving to different parts of a spinning spacecraft or when changing the spin rate might induce symptoms of disequilibrium.

A comprehensive program is required to (1) determine the gravity threshold required to reverse or prevent the deleterious effects of microgravity and (2) evaluate the effects of centrifugation on behavior and/or sensorimotor function. Part of the required research could be accomplished by using human surrogates, including nonhuman primates, on a dedicated centrifuge in low Earth orbit. Studies of human responses to spinning will require a centrifuge of sufficient dimension to accommodate humans. An alternative strategy would be to investigate the use of rotating tethered spacecraft[16] to provide artificial gravity. It is possible that the detrimental vestibular effects of spinning can be eliminated if the tethers are sufficiently long.

Even assuming an optimistic schedule for lunar operations or space station activation, the relevant life-sciences knowledge developed from them will probably not be available before the beginning of the second decade of the 21st century. This implies a substantial technical risk in any program of Mars exploration that relies on a comprehensive solution to problems of human adaptation to microgravity. The prudent alternative is to carry forward, during conceptual design phases, alternatives providing for artificial gravity (as recommended in a National Research Council report[17]) during the cruise flight phase, and possibly in Mars orbit as well. If satisfactory countermeasures are confidently identified during a vigorous and rigorous program of orbital life-sciences research, this alternative design path can be abandoned. Conversely, if an effective artificial-gravity system is developed, research on countermeasures will become less urgent.

The design, construction, and operation of rotating spacecraft may pose formidable technical challenges. Nonetheless, all investments in the program will otherwise be hostage to a favorable outcome in the human adaptation issue. In the view of CHEX, the Synthesis Group's report erred ab initio in discarding consideration of artificial-gravity scenarios in its four architectures. Indeed, the provision of artificial gravity may well prove to be an architectural variable of more fundamental importance than the thematic differences between alternative mission emphases presented in the report of the Synthesis Group.

CARDIOVASCULAR AND PULMONARY FUNCTION

The redistribution of intravascular fluid toward the head is one consequence of exposure to a microgravity environment.[18] This shift has not impaired astronauts' cardiovascular and cardiopulmonary function during the relatively short periods of exposure to microgravity experienced thus far. It has, however, caused clinically significant dysfunction following return to Earth. This dysfunction manifests itself as an orthostatic intolerance and decreased capacity for exercise. Full recovery appears to occur rapidly (within 2 to 5 days) following short flights but can take as long as 30 days following long flights. The potential exists for permanent impairment following prolonged adaptation to microgravity. Both acute and longer-term problems could occur upon landing on Mars, since its gravity is only about three-eighths that of Earth's. With limited health support available, reduced cardiovascular function could threaten the success of crew activities on Mars.

Microgravity leads to a reduction in plasma volume that also contributes to orthostatic and exercise intolerance upon return to Earth. When the blood volume in the chest and head increases, the kidneys excrete more fluid. Another factor contributing somewhat to orthostatic hypotension and reduction in exercise performance is a decrease in total red blood cell mass. When exposures to microgravity are brief, both of these effects are reversible.

Atrial and ventricular rhythm disturbances have occurred with significant frequency in both astronauts and cosmonauts and thus require attention. Particular examples include the following:

1. One cosmonaut was prematurely returned from Mir because of a refractory atrial rhythm disturbance.

2. Apollo 15's lunar module pilot sustained premature ventricular contractions (PVCs) with some episodes of bigeminy; 60 hours later he also had premature atrial contractions (PACs). Apollo 15's commander also sustained a run of PVCs.

3. The crew of Skylab 3 showed occasional PVCs and ectopic supraventricular contractions.

4. Atrioventricular block of brief duration has been observed in several crew members upon release of lower-body negative pressure before reentry.

5. PACs have been observed in several astronauts during extra-vehicular activity.

The mechanisms for these effects remain unknown but could be related to shifts in intravascular volume and ensuing perturbations of regulatory hormones. The significance of these effects is also unknown but *could* be a prelude to more severe problems.

Further studies of the response of humans and animals to changes in gravitational force are essential to complete our understanding of the mechanisms responsible for cardiovascular and pulmonary deconditioning in space. Questions about the reversibility of deconditioning can be answered only by careful studies of animals and eventually humans, during and after prolonged exposure to microgravity. Adequate experimental controls require a centrifuge designed to accommodate primates.

Specific high-priority areas of cardiovascular investigation include:

1. The role of exercise and physical fitness before, during, and after flight;
2. Countermeasures against cardiovascular dysfunction during flights and rehabilitation after long flights;
3. Validation of ground-based models of microgravity for short-term and long-term studies; and
4. Characterization of drug pharmacodynamics in microgravity.

It is necessary to study the effects of long-term spaceflight on:

1. Cardiodynamics (e.g., cardiac output, chamber pressures and dimensions, and performance);
2. Cardiac rhythm (as shown by electrocardiograms taken at rest and during maximum exercise);
3. Hormone release and metabolism (e.g., of antidiuretic hormone, atrial natriuretic peptide, and aldosterone);
4. Baroreceptor function (neural regulation of blood pressure);
5. Peripheral resistance (resistance offered to blood flow through the circulatory system); and
6. Pressures, degree of tone, and capacitance of the venous system.

Ventilation and blood flow to the different regions of the lung are affected by gravity and so will obviously be affected by microgravity. To quantify these effects, studies of the rate and depth of respiration, the component lung volumes, air flow, gas exchange, and pulmonary pressures at 1 g and at different levels of microgravity are necessary.

Another topic needing attention is potential effects of the space environment on cardiovascular and pulmonary physiology when modified by disease processes or pharmacological agents.

BEHAVIOR, PERFORMANCE, AND HUMAN FACTORS

Empirical evidence suggests that the performance of crews composed of competent, highly trained individuals is critically determined by psychological and social factors.[19] Moreover, psychosocial considerations necessarily assume greater importance when people are confined in isolated and inescapable environments. Reports from both cosmonauts and astronauts confirm the importance of psychological factors during long-duration missions. Despite awareness of the importance of these issues, systematic research into the determinants of human performance and adaptation under these conditions has received only minimal support. Only limited progress has been made since publication in 1987 of the Committee on Space Biology and Medicine research strategy, which included a chapter on human behavior.

Because of the limited number and duration of American spaceflights, systematic research in this field could be conducted in analog environments such as polar stations, undersea habitats, and aviation settings. However, generalizing the results of research in such analogs has its limitations. Nevertheless, available data strongly indicate that focused research on small groups in confined quarters may result in practical knowledge that could reduce the incidence of interpersonal conflict and psychological problems. The utility of such data should be even greater when groups work for prolonged periods in isolation and when experimental interventions can be conducted under controlled conditions.

The psychological factors relevant to the success of a mission can be organized into three domains: individual, group, and environmental. More basic research is urgently needed in each area. In addition to investigations in analog environments on Earth, the psychological determinants of current space operations, even short-duration shuttle missions, need more intensive study. Any single investigation, however, will lack features of a Mars mission such as the microgravity environment, exposure to radiation, mission duration, and lack of escape capability. Nevertheless, the aggregate findings from many such studies should provide important guidelines for the planning and conduct of very long missions.

Individual Factors

Just as technical competence is a prerequisite for task fulfillment, so also will the personality and motivation of each crew member critically influence the success of long-duration space missions. Efforts must be directed toward determining psychological profiles associated with performance and adjustment under conditions of prolonged isolation. Psychological selection strategies must be refined to focus not on screening out those candidates showing evidence of psychopathology but rather on selecting those candidates with optimal attributes.

Disruption of normal circadian (i.e., 24-hour) rhythms is another important factor to consider when planning long spaceflights. If unchecked, such disruption can lead to serious perturbations in human performance and productivity, with both psychological and physical consequences. Problems arising during exploration missions may be particularly severe since these rhythms appear to be disrupted by microgravity and/or high stress. Studies are needed to determine the optimal environmental conditions necessary to create the sense of normal circadian rhythms within the body during long-duration space missions.

Group Factors

Even the most technically competent and highly motivated individuals do not necessarily perform effectively and harmoniously when sequestered for prolonged periods in a confined environment. Moreover, the effects of seclusion can be exacerbated if escape is impossible. Improved methods are necessary for selecting and training teams so that they can sustain high levels of motivation, work quality, and interpersonal relationships. Training techniques developed to improve leadership, crew coordination, decision making, and conflict resolution in civil- and military-aviation settings need to be refined and validated in the space environment.

Environmental Factors

On long spaceflights, the crew's psychological environment is no less important than its physical environment. Additional research in operational, analog settings is required to determine the best social organization for human exploration missions. Issues central to crew effectiveness include:

1. How to organize daily activities to maximize performance and satisfaction (e.g., by providing meaningful, intellectually challenging work and enjoyable leisure activities) and to avoid boredom;

2. How to establish levels of automation that will balance efficient operations against operator control and satisfaction; and

3. How to establish an optimal division of responsibility between ground and space components to provide appropriate mission control while maintaining an efficient, cooperative relationship. Since crew safety is of paramount importance, the spacecraft commander must be vested with the final authority in all questions relating to the crew's health and welfare.

The design of the physical environment for long-duration missions should be based on research into requirements for privacy, habitability, and social interaction. A balance is necessary between engineering constraints and the requirements for harmonious group living over extended periods. In addition, the characteristics of the physical environment and the scheduling of work, leisure, and sleep cycles should minimize disruption of normal circadian functions. Many of these environmental and organizational issues could be profitably investigated in polar research stations and undersea habitats.

BIOLOGICAL ISSUES

The biological aspects of missions to Mars fall into two categories: those related to human well-being and those related only to exobiology. These overlap if a crew member is infected by a putative martian microorganism or if such organisms are returned to Earth. Although the chance is small that organisms, pathogenic or otherwise, exist on Mars today, public and legal concerns dictate close attention to this issue.

The protocols for the preparation of Mars-bound craft or the handling of martian samples returned to Earth will depend both on the relevant planetary protection regulations promulgated by the Committtee on Space Research (COSPAR) and on public perception of the risks. The latter arises now much more stridently than it did in the past when the issues of forward and back contamination were first raised. Existing COSPAR regulations (currently under review) may require that landers be sterilized to prevent the introduction of terrestrial organisms to the martian environment.[20] The Viking spacecraft, for example, were decontaminated by a combination of presterilizing components and dry-heating the assembled landers prior to launch. Although these procedures were time consuming and extremely expensive, it may be required that they be applied to future robotic missions. Similarly, there is no question that rigorous procedures will be required for handling samples returned to Earth by robotic missions. A recent study[21] has concluded that the question of forward contamination by robotic missions is an issue only for those that include life-detection experiments, where the concern is contamination of the experiment. It would, however, be virtually impossible to avoid forward contamination of Mars or back contamination of Earth from human exploration.

Using the return flight as an incubation period and the crew as guinea pigs (as has been suggested[22]) is not a solution to back contamination on human missions. Would the whole mission be risked if an unanticipated contamination occurred? How would the cause of an infection be known with enough certainty to justify destroying the returning spacecraft before it entered Earth's atmosphere? The whole spacecraft, not only the astronauts, would be contaminated. In addition, infection might not be the only risk. A returning organism could possibly cause some long-term changes in our environment, perhaps remaining undetected for a while. Although such an event may be judged to have a very low probability, a convincing case that prudence has been exercised will have to be made to the public.

The scientific requirements relating to planetary protection and the assessment of the possibility of health-threatening microorganisms include:

1. How to detect the presence of indigenous microorganisms (potential pathogens) and their activities in samples returned to Earth prior to a human visit to Mars. A corollary is how to certify the biological safety of samples returned to Earth and of potential sites for human habitation. Simple culture experiments are insufficient because some organisms (e.g., the cholera-causing pathogen *Vibrio cholerae*) are not culturable using standard microbiological techniques. In fact, there is no unbiased assay to enable detection of even terrestrial microorganisms present at low concentrations.

2. How to detect potential pathogens during residence on Mars. The need for such detection may arise as novel habitats are encountered or as humans make use of martian resources such as water.

3. How to treat and handle an explorer in the highly unlikely event of infection by a martian life form.

4. How to monitor the fate and impact of terrestrial microorganisms unavoidably transported to Mars by vehicles or humans.

Addressing these issues will involve investigations of Mars-like environments on Earth as well as laboratory studies to develop the necessary tests, procedures, and protocols.

NOTES AND REFERENCES

1. Space Science Board, *Space Science in the Twenty-First Century: Imperatives for the Decades 1995 to 2015: Overview*, National Academy Press, Washington, D.C., 1988, pp. 67-68. Also see Advisory Committee on the Future of the U.S. Space Program, *Report of the Advisory Committee on the Future of the U.S. Space Program* (the "Augustine report"), U.S. Government Printing Office, Washington, D.C., 1990, p. 6.

2. Synthesis Group, *America at the Threshold*, Report of the Synthesis Group on America's Space Exploration Initiative, U.S. Government Printing Office, Washington, D.C., 1991, p. 24.

3. See Ref. 2, p. 27.

4. NASA, *Report of the 90-Day Study on Human Exploration of the Moon and Mars*, NASA, Washington, D.C., November 1989, p. 6-2.

5. See Ref. 4, p. 6-3.

6. International Commission on Radiological Protection, Recommendations of the International Commission on Radiological Protection, ICRP Publication 60, *Annals of the ICRP* 21(1-3):1-201, 1991.

The equivalent dose in tissue, H_T, is given by the summation:

$$H_T = \Sigma_R W_R \cdot D_{T,R}$$

where $D_{T,R}$ is the absorbed dose averaged over the tissue or organ T due to radiation R. The weighting factor, W_R, is selected for the type and energy of the radiation incident on the body. Its value is based on the relative biological effectiveness (RBE) for the radiation in *inducing stochastic effects (that is, cancer and genetic damage) at low doses*. It is vitally important to remember that these effects of long-term exposure to relatively low doses of radiation are in addition to any deterministic effects (the physical damage to tissues) likely to result from large doses of ionizing radiation.

7. The sievert is the unit of dose equivalent and is equal to the absorbed dose (in grays) multiplied by the quality factor. One gray is an absorbed dose of 1 joule per kilogram. One sievert = 100 rem. One gray = 100 rad.

8. J.H. Adams, Jr., G.D. Badhwar, R.A. Mewaldt, B. Mitra, P.M. O'Neill, J.F. Ormes, P.W. Stemwedel, and R.E. Streitmatter, "The Absolute Spectra of Galactic Cosmic Rays at Solar Minimum and Their Implications for Manned Space Flight," *Galactic Cosmic Radiation Constraints on Space Exploration*, NRL Publication 209-4154, Naval Research Laboratory, Washington, D.C., November 1991.

9. J.R. Letaw, R. Silberberg, C.H. Tsao, and E.V. Benton, *Advances in Space Research*, Vol. 9, No. 10, p. 257, 1989.

10. J.H. Adams, Jr., and A. Gelman, "The Effects of Solar Flares on Single Event Upset Rates," *IEEE Transactions on Nuclear Science*, NS-31, pp. 1212-1216, 1984.

11. National Council on Radiation Protection and Measurements, *Guidance on Radiation Received in Space Activities*, NCRP Report No. 98, National Council on Radiation Protection and Measurements, Bethesda, Maryland, 1989, p. 25.

12. Space Studies Board, Letter to James D. Watkins and Daniel J. Goldin, August 20, 1992, Washington, D.C., 1992.

13. Space Science Board, *A Strategy for Space Biology and Medical Sciences for the 1980s and 1990s*, National Academy Press, Washington, D.C., 1987, p. xiii.

14. See Ref. 13, p. ix.

15. See Ref. 13, Chapter 5.

16. L.G. Lemke, "An Artificial Gravity Research Facility for Life Sciences," *Proceedings of the 18th Intersociety Conference on Environmental Systems*, American Institute of Aeronautics and Astronautics, 1988. For more information on tethers, see P.A. Penzo and P.W. Ammann (eds.), *Tethers in Space Handbook*, Second Edition, NASA, Washington, D.C., 1989.

17. Committee on Human Exploration of Space, *Human Exploration of Space: A Review of NASA's 90-Day Study and Alternatives*, National Academy Press, Washington, D.C., 1990, p. xi.

18. See Ref. 13, Chapter 6.

19. See Ref. 13, Chapter 11.

20. Space Science Board, *Recommendations on Quarantine Policy for Mars, Jupiter, Saturn, Uranus, Neptune, and Titan*, National Academy of Sciences, Washington, D.C., 1978. See also Space Studies Board, *Biological Contamination of Mars: Issues and Recommendations*, Task Group on Planetary Protection, National Academy Press, Washington, D.C., 1992.

21. Space Studies Board, *Biological Contamination of Mars: Issues and Recommendations*, National Academy Press, Washington, D.C., 1992.

22. See Ref. 2, p. 77.

3

Research for Mission Optimization

This chapter describes several issues that are relevant to the health and well-being of humans but that appear, at present, to represent less critical threats to the lives of astronauts than those discussed in the previous chapter. They are, however, no less important as related to optimum human performance during exploration missions. In addition, increased knowledge of the physical aspects of the Moon and Mars is required to ensure that human explorers perform efficiently. As new information accumulates, and as implementation decisions are made, the significance of any or all of the areas where research is needed to ensure mission optimization could increase to the point that they become critical issues.

SENSORIMOTOR INTEGRATION

Changes in the gravito-inertial environment during a space mission may lead to disturbances of sensorimotor function.[1] The consequences may include impaired spatial orientation, instability of position and gaze, and motion sickness. Fortunately these problems are of short duration because the central nervous system adapts to those changes within a few days *provided a constant environment is maintained.* There are, however, two caveats to this assessment of relative risk. First, gravito-inertial changes occur at the most critical times during a mission: takeoff and landing. Second, the crew of a spinning spacecraft (possibly used to counter the problems associated with prolonged exposure to microgravity) might suffer repeated changes in their gravito-inertial environment when moving to different parts of the craft or if the spin rate is changed. Fortunately, no known long-term health risks are associated with sensorimotor adaptation to microgravity.

Although both the National Institutes of Health and NASA are studying vestibular function and its interaction with other sensorimotor modalities, the etiology of motion sickness in general, and space adaptation sickness in particular, is still not known. The extent to which adaptive responses can be shaped or overridden by appropriate training in sensorimotor strategies is also unknown. Studies of vestibular function and its neuronal substrates in appropriate animal models are needed both on the ground and in a microgravity environment. Parallel studies of human sensorimotor performance in both environments must also be pursued.

IMMUNOLOGY

Can the immune system be damaged by spaceflight? This possibility stems from observations of abnormalities in the two major types of human lymphocytes, T-cells and B-cells, and in other white blood cells on the Spacelab D-1 mission. A reduction of function and disordered morphology of T-cells have been detected on some other flights. Moreover, changes in rat immunity have been observed on spaceflights conducted by the former Soviet Union.

Serious infections in humans during spaceflights are rare. Thus, there have been no opportunities to systematically assess the capacity of humans or other mammals to contain and eradicate infections by various types of terrestrial microbes while in space. The potentially devastating consequences of any immune dysfunction, particularly on long-duration flights, indicate the urgent need for further studies. The possible defects already identified in lymphocytes and also other elements of immunity vital to specific and adaptive defense mechanisms in humans need to be examined.

The potential effects of spaceflight on normal human immunity must be judged in terms of the antibody responses and reactions of lymphocytes, macrophages, and other white blood cells to different types of antigens. The most common antigens on Earth are proteins, carbohydrates, and complex lipids. These are presented to the immune system in soluble form and as a part of cells or other complex structures. The studies of responses to antigens in space should use both intact microbes, to mimic infections, and soluble purified proteins and carbohydrates, to simulate simple vaccines.

A vital aspect of immunity is a memory of exposure to antigens. Thus, comprehensive studies should encompass both new and previously encountered antigens of each major chemical class and physical form. This diversity of experimental challenges is critical for assessment of immunity in space, because the variety and intensity of antigen challenges to the immune system will be substantially different from that experienced on Earth: the unique closed environment imposed by the spacecraft offers significantly decreased opportunities for the constant bombardment by new antigens encountered on Earth. The potential problem is that the immune system could become atrophic and render an individual more vulnerable to infection (especially if sufficiently rigorous measures are not taken to control microfloral contamination of the spacecraft).

If the T-cell defects are confirmed, then their effects should be delineated in relation to four factors:

1. The differences in responses to antigens and broader cell stimuli called mitogens;
2. Abnormalities in subsets of regulatory T-cells, which help or suppress activities of other immune cells;
3. The roles of diverse immune-cell-derived regulatory proteins called cytokines, which direct T-cell proliferation and functions; and
4. The functions of macrophages and other accessory white blood cells responsible for presenting antigens specifically to T-cells.

Effector systems, which eliminate toxins and kill microbes targeted by antibodies, such as white blood cells of the granulocyte series and serum proteins called complement factors, also should be assessed functionally. Some in vivo studies are required to detect and understand any deficiencies or excesses in integrated human immune responses.

The critical need for controlled variable-gravity studies cannot be overemphasized. Only such studies will produce data useful in identifying specific mechanisms, perceiving the impact of any immune system abnormalities on other systems, and providing clinical guidelines for preventing and countering any defects in human immune defenses.

The closed environment of the spacecraft may encompass a variety of living organisms (e.g., humans, animals, and plants), many types of energy-using equipment, and a wide variety of materials. The effluent from these multiple sources will contain microflora, gases (e.g., oxygen, carbon dioxide, and methane), and other chemical contaminants that must be collected and either disposed of or channeled through the life support system. The accumulation of colonies of microflora, pockets of gases, or dispersed trace chemicals could jeopardize the health of a crew and interfere with the success of a mission.[2] At this time we do not have adequate information to assess how microbial and immunological problems would affect humans during extended spaceflight.

DEVELOPMENTAL BIOLOGY

A major scientific goal of studying developmental biology[3] in space is to "evaluate the capacity of diverse organisms, both plant and animal, to undergo normal development from fertilization through the subsequent formation of gametes under conditions of the space environment."[4]

Plants are key to the entire biological system that has developed on Earth. Thus, it is essential to understand the effects of gravity and its absence in order to grow plants in space for food or for use in life support systems (see next section). A considerable amount of scientific literature already exists on the biology of plants in space. However, most studies have not dealt with general questions about plant growth but, rather, have addressed the orientation and motion of roots and shoots or have focused on plant hormones and events associated with normal and gravi-stimulated cell and organ growth. Our understanding of plant signal transduction is scant and may well be enhanced by using models based on animal work. Such constituents as G-proteins, phosphoinositides, actin, and calmodulin also occur in plant cells and may have active roles. The increasing applicability of techniques of molecular biology to problems in plant growth and development will be useful in attempts to understand the responses of plants to the space environment and in developing breeding programs designed to increase plant performance in microgravity environments.

A major question is whether plants are capable of producing multiple generations in microgravity. The definitive space experiment is to observe a plant's life cycle from seed to seed to seed. The first generation of "on-orbit" seeds could have ground-born flowers upon germination, and thus produce seeds with ground-born tissues, since seed has maternal material in it. These seeds, however, would produce flowers exposed only to microgravity. Thus, their offspring, the third generation of seeds, would be entirely free of any prior terrestrial gravitational influence.

Another important question is whether microgravity affects the single cell or if some plant cells acclimate to gravity deprivation. Some space-based studies suggest that chromosome behavior is fundamentally changed in microgravity. Should this be the case, the consequences and their implications for cell development must be determined.

The lack of thermal convection in the microgravity environment may affect short- and long-distance transport phenomena in plants. For example, the function of cell membranes, the pathways for ion uptake and nutrient absorption, plant-water relations, and the transport of organic and inorganic molecules must be investigated to determine whether any of these is affected by microgravity. For example, is the plant-supporting structure of lignin and cellulose modified in space in ways analogous to the loss of bone density?

LIFE SUPPORT SYSTEMS

Closely related to the question of plant growth in space is the feasibility of a closed-loop life support system (CLLSS). CLLSSs are integrated self-sustaining systems capable of providing potable water, a breathable atmosphere, and ultimately, food for astronauts on long-duration missions. Some such systems may be able to operate in a small enough volume to be practical in a space vehicle, while larger systems could be deployed at lunar and martian outposts. Although it is not yet clear if the initial phases of the human exploration of Mars demand a CLLSS, it is certain that without one, long-term missions will require either vast amounts of on-board stores or access to prepositioned supplies. Thus, an effective and reliable CLLSS, even if limited to generating air and water from crew waste, would greatly simplify the logistics of long-duration missions.

While a first-generation CLLSS would recycle only air and water, more advanced versions would be highly integrated subsystems for plant growth, food processing, and waste management. We have very little data on the operation of individual system components under realistic conditions. A small amount of information has been gathered on the performance of a few arbitrarily chosen plant species in open growth chambers. In addition, some encouraging, but still tentative, experiments have been initiated on plant growth in closed environments. Virtually nothing seems to have been done with respect to microbial and other systems of waste recycling, soil microbes and other microflora, or pathogen control. Nor have any of the food-processing technologies for converting biomass into palatable human nutrients been developed.

Green plants are critical components of even the simplest CLLSS. They can fix carbon dioxide, produce food and oxygen, and purify water. However, as noted in the previous section, we do not yet know if plants will grow in space well enough to support a CLLSS for significant periods of time. A major scientific goal is simply to grow plants in space for extended periods of time—over several life cycles—while carefully monitoring their performance. This goal is related to the more general need, outlined in the previous section, to investigate how diverse organisms undergo development in the space environment. For development of a CLLSS, this overall scientific goal assumes immediate practical importance. As we have already seen, processes such as reproductive development, fluid transport, and photosynthetic gas exchange may be adversely affected in low-gravity and microgravity environments. Even small effects may have serious consequences when performance is integrated over long time periods.

Many other components of a CLLSS must also receive attention. Diverse plant, animal, and microbial species must be evaluated, environmental parameters optimized, and procedures developed for food processing and for recycling liquid and solid waste materials. In many cases, we do not know enough to produce a suitably sized CLLSS on the ground, much less in space. Obtaining the required scientific knowledge and engineering experience will require extensive experimentation under actual conditions in space.

MICROMETEOROID FLUX ON THE MOON

Long-duration activities on the surface of the Moon increase the potential risk of experiencing lethal impacts by micrometeoroids. The use of average collisional fluxes may give a false sense of security as excursion times outside protective habitats increase. The occurrence of periodic terrestrial meteor showers related to comets is well known. Recent reanalysis of lunar seismic data reveals that lunar impacts are neither temporally nor spatially random. Moreover, not all observed meteoroid showers on the Moon correlate with known terrestrial meteor showers.

The potential dangers meteoroids pose to a long-duration presence on the Moon are twofold. First, there is an increased risk of direct hits during peak activity. Second, there is a risk of high-velocity impacts from secondary and ricocheting debris. The potential for lethal damage depends on the actual flux, the size distribution of the impactors, and the effect of spatially clustered impacts. These unknowns need to be studied over a sufficiently long period not only to assess the short-term risks (day to month), but also to recognize annual events and possible catastrophic swarms during orbital passage of newly discovered comets.

Lunar seismometers have proven their usefulness as meteoroid impact detectors. Establishing a seismic network on the Moon to characterize the flux, size distribution, spatial clustering, and possible directional anisotropies of impacts over a multiyear period is essential to evaluating the hazards posed to astronauts by meteoroids. The potential dangers of unexpected meteoroid storms can be assessed through continued monitoring and evaluation of newly discovered comets. Experience gained from seismic monitoring of small impactors will be important for assessing risks over even greater durations en route to, and in orbit around, Mars.

SURFACE AND SUBSURFACE PROPERTIES

Humans exploring the Moon and Mars will require knowledge about their proposed landing sites not only to ensure a safe touchdown and subsequent departure, but also to identify regions of potentially high scientific interest. Prime questions to be answered for candidate sites involve the mechanical properties of the landing zone and the surrounding terrain to be explored and sampled. Size distributions of rocks at potential landing sites are required for three reasons: first, to ensure sufficient clearance for the landing vehicle; second, to allow reasonable leveling of the lander; and third, to certify that the terrain is sufficiently benign to be traversed by astronauts on foot and with rovers to carry out mission objectives. Of equal importance is a priori knowledge of the mechanical or bearing strength of the surface, particularly at the precise landing site but also over the region to be explored by the astronauts.

The distribution of rock size can be obtained by precursor flights using remote sensing and in situ robotic exploration. Imaging with a resolution of less than 1 meter is necessary for selecting the landing sites themselves. Information on bearing strength is more difficult to obtain remotely. Significant estimates can be made of the near-surface soil densities using radar reflection and microwave emission techniques. Robotic landers may be required to achieve sufficient confidence to certify sites for human landings unless the areas selected are familiar (e.g., Apollo or Viking sites or demonstrably similar ones).

In addition to rocks, the lunar surface is blanketed with unconsolidated debris generated by meteoroid impacts. This material, called regolith or soil, contains broken mineral and rock fragments, impact-produced glasses, and rocky glass-bonded aggregates. On average, about 20% of the regolith is composed of particles smaller than 20 microns in size. These properties, coupled with the hard lunar vacuum (10^{-12} to 10^{-14} torr), make the regolith extremely abrasive. This will affect the longevity of all moving parts it comes in contact with. To make matters worse, regolith tends to cling to surfaces, leading to additional wear and tear on mechanisms such as gears, habitat airlocks, and spacesuit joints. Further in situ and remote sensing of the lunar surface and subsurface, together with studies of the abrasive and adhesive properties of lunar soil under hard vacuum conditions in terrestrial laboratories, will help in designing equipment to operate on the Moon's surface. Large-scale simulation facilities might also be needed to conduct long-duration, full-scale tests on engineering equipment and transport vehicles.

The nature of the lunar subsurface at depths of 1 to 10 meters is poorly known. Although the size distributions of surface blocks in the regolith are known for typical mare and highland regions, there is little knowledge of how these distributions may change with depth. In most regions, bedrock occurs at depths of just a few meters, but the nature of its interface with overlying fragmental debris is unknown. Moreover, subsurface discontinuities, including interbedded lava flows, bedrock ledges, and voids, may pose additional hazards to landing craft, rovers, and excavation equipment. The elimination of such hazards may require active seismic imaging.

Like the lunar regolith, the martian surface material may also be hazardous, but for different reasons. Existing data show that it contains highly reactive components in sufficient concentration to have oxidized the organic compounds used in one of the Viking life-sciences experiments. Such compounds may perhaps be responsible for the complete absence of any organic compounds in samples examined by Viking's gas chromatograph/mass spectrometer. Toxicity analysis could probably be carried out by a precursor robotic mission and might not require the analysis of martian material in terrestrial laboratories.

Based on current knowledge, the oxidizing material is likely to be associated with fine, windblown, particulate material. Thus, specific precautions against this dust will have to be built into the airlock system on a lander. Moreover, spacesuits will have to be decontaminated as astronauts reenter the lander after completing extra-vehicular activities. Perhaps the spacecraft itself will have to be "cleaned" prior to its return to Earth.

The data required to certify landing sites for safety may be highly desirable for other purposes such as planning surface construction, instrument installation, and the layout of extended surface traverses. Construction, prospecting, and mining operations will require subsurface sampling around the landing point. This can be carried out by the astronauts if the site has been selected on the basis of good information from precursor flights. That is, good measurements of surface rock distributions can be used to infer the subsurface geology. For Mars, such information is particularly critical because broad regions of the planet were not emplaced as primary geologic units, but, rather, have undergone episodic resurfacing tied to atmosphere-surface interactions. Astronauts can locate regions free of subsurface hazards for construction and mining using seismic and electromagnetic sounding devices on their rover.

The need for some of these data could be partly alleviated through the use of a robust and forgiving design for excavation and construction equipment. For example, if the capability to efficiently crush and remove rock is a requirement for a lunar bulldozer, the need for knowledge of the sizes and locations of subsurface boulders is diminished.

POTENTIAL MARTIAN HAZARDS

Potential hazards posed by martian weather and climate, volcanic and seismic activity, and a number of other factors need to be considered in the context of concern for astronaut safety and the major investment of resources in any program of human exploration. A mission failure due to lack of adequate assessment of all plausible and sensible potential hazards, however unlikely, would be inexcusable. Following appropriate studies, some of the potential hazards may be realized; others may turn out to be either non-existent or of such low probability that they can be dismissed.

Severe martian weather (such as dust storms, dust devils, and other vortices) may pose hazards to man-made structures or to field operations. Data on near-surface winds, including local wind shear and vorticity, are available only for the two Viking lander sites. Winds may affect descent vehicles by posing a hazard to, for example, parachute deployment or the spacecraft's ability to land precisely at a desired site. Ascent vehicles may also be affected by strong wind shears or turbulence. Variations of atmospheric density with local time, with solar activity, and with variations in the lower atmosphere (e.g., dust storms) may affect the operations and lifetimes of near-Mars support spacecraft, such as site-reconnaissance orbiters and communications satellites. Long-term meteorological measurements of temperature, pressure, wind velocity, and dustiness from orbit and at a variety of surface sites are required to assess these hazards. The current Mars Observer mission is directly relevant to this need.

Large dust devils and clouds associated with local storms have been observed. Although dust storms may occur in any season, one or more may grow to regional and, on occasion, even global scale during southern spring and summer. Dust storms reduce surface visibility and insolation, thus affecting, for instance, the efficiency of solar cells. Moreover, the movement of sand-sized particles near the surface may pit, scratch, and erode surfaces, and may foul joints. Continued remote sensing of the martian atmosphere will help define this hazard.

As is the case on the Moon and in free space, components of solar radiation reaching the surface of Mars may pose hazards to field workers and equipment (e.g., ultraviolet degradation of plastic material). Unlike the lunar surface and space, however, the total flux and the spectral distribution will change with variations in atmospheric aerosols and the seasons.

Information on the diurnal and seasonal variation of atmospheric temperature, density, and wind speeds is needed to design a martian outpost. Other factors such as local and regional topography can present additional hazards (e.g., strong winds on steep slopes or in canyons, or regions of local fogs). Certification of landing and base sites in regions of large interannual variability (mainly at mid and high latitudes) may require observations spanning several martian years or longer to characterize the complete range of conditions likely to be experienced.

Practically nothing is known about electric fields on Mars. The presence of moving dust particles in an atmosphere nearly as dry as Earth's stratosphere, however, could produce significant electrostatic charging. Besides being a nuisance (e.g., fine dust clinging to optical surfaces), such charging and discharging could severely affect crucial electrical equipment, such as computers. Large discharges—such as lightning—may also occur.

Although the hazard posed by meteorites falling on Mars is small, the impact flux could range from a nominal lunar value to one larger by as much as an order of magnitude. The circum-martian meteoroid flux could be determined by a spacecraft akin to NASA's Long Duration Exposure Facility in orbit around Mars, by detectors of meteors passing through the martian atmosphere, and by seismic networks on the martian moons.

The long-term safety of a martian outpost also requires assessment of the hazards due to seismic or volcanic activity. Insufficient data currently exist to make confident statements about martian seismicity. Volcanic activity has been widespread on Mars in the past. We do not know, however, if there has been any recent volcanism or if near-surface thermal activity or magma chambers exist. A network of seismometers and heat-flow measuring devices could provide the information to measure current activity. Other geologic hazards, including slides and slope failures, need to be assessed.

Areas of scientific interest in potentially dangerous locations, such as deep martian canyons or close to known volcanic vents, may require precursor visits by robot landers or rovers. Such sites may be especially important in deciphering the history of Mars, particular the role played by liquid water in both geological and biological contexts.

AEROBRAKING AT MARS

Aerobraking, or aerocapture, is a technique using atmospheric drag to reduce a space vehicle's orbital energy. It can thus cut down on the amount of propellant needed to achieve orbital insertion. Indeed, aerocapture may significantly reduce (perhaps by a factor of three or more) the mass that must be delivered into Earth orbit for a Mars exploration mission. Aerocapture could be critical to the feasibility of such a mission, and a proper understanding of the atmospheric structure of Mars and its variability should be considered part of the enabling science for such a mission.

Successful aerobraking requires a detailed knowledge of not only the mean density structure of the martian atmosphere but also its temporal and spatial variations. The Viking 1 and 2 landers, for example, measured vertical density profiles differing by more than 20% as they descended from an altitude of 100 kilometers to the surface. Most of the atmospheric variations at aerobraking altitudes on Mars (20 to 70 kilometers) are due to gravity waves. These are thought to be generated by thermal tides and by high-speed winds flowing over surface topography.

Further understanding of the statistics of density variations in the martian atmosphere is required before human landings using aerobraking are attempted. NASA's Mars Observer mission should answer many of the outstanding questions on this issue. However, a longer mission (with greater seasonal coverage) and some in situ measurements of the atmosphere will be required to calibrate remote observations. A better understanding of the temporal and spatial variations of atmospheric dust is also needed and should be obtained either from direct atmospheric measurements or by ground-based opacity observations. These concerns are addressed in considerable detail in a recent report.[5] This NASA document likewise concludes that mission safety requirements lead to a significant need for understanding the statistical behavior of the martian atmosphere. Remote spacecraft monitoring of atmospheric properties should be carried out both before and during the arrival of humans at Mars.

MICROGRAVITY SCIENCE AND TECHNOLOGY

Human exploration will require more understanding of fluid flow and transport under reduced (and sometimes increased) gravity conditions. In order to support extended space travel, we must know more about the processing of materials, thermal management, and the handling of fluids. Microgravity studies must be viewed as more than the advancement of science and technology for its own sake or as a means to obtaining potential benefits for society on Earth; these studies are essential to the advancement of spaceflight.

Many examples of challenges associated with a modified gravity field can be found: producing needed materials from available raw materials; washing and drying of clothing, equipment, humans, and animals; handling of hazardous and obnoxious wastes; improving and ensuring spacecraft fire safety; and achieving temperature control for humans, animals, plants, and electronics. The challenges occur predominantly in the life support areas but extend well beyond them. For example, modern electronics are becoming so compact that, in the near future, volumetric heat-generation rates are expected to rival those values for controlled nuclear fission. Also, there is overlap with the life sciences since fluid transport is essential to life itself, as, for example, the transport of liquid from the roots to the leaves of plants.

There is a strong need to address the underlying science as well as the technology. The relevant technology for related Earth-gravity-level processes is often based on empirical methodology. Therefore, engineering extrapolations cannot be readily made.

EXOBIOLOGY ISSUES

While there may be little chance that life exists on Mars today, this may not always have been the case. Thus, many of the science requirements relating to exobiological exploration of Mars revolve around technologies for detecting and analyzing fossil organisms or the chemical precursors to life. Closely related is the question of the history and present occurrence of liquid water and ice on Mars. Some specific questions include:

1. How to detect indigenous martian microorganisms and assess their biological activities;
2. How to recognize and analyze fossil remains of such indigenous microorganisms;
3. How to search for the presence of chemicals that might relate to past activities of life forms or that might relate to prebiotic chemistry;
4. Where to seek evidence for past life or prebiotic chemistry; and
5. How to detect the current, and understand the past, distribution of liquid water and ice.

Beyond laboratory studies, answering these questions will involve acquiring a more detailed knowledge of Mars and its history. The location of ancient lake beds and of possible wind- and water-emplaced sediments will surely play a major role in selecting martian sites of interest to exobiologists.

The development of new organic analysis instrumentation with perhaps a 1000-fold improvement in sensitivity over the Viking mass spectrometer is likely to be needed. This needs to be coupled with a flexible "wet" chemistry input. If we are to adequately investigate the possible prehistory of biology on Mars, we need to answer whether or not there are any organic compounds of either abiogenic or biogenic origin on the surface or below the surface. Determining the ratios of different stereoisomers of amino acids will help distinguish between those of biogenic or abiogenic origin.

RESOURCE UTILIZATION

Long-term human exploration of Mars may require or greatly benefit from landing sites in close proximity to exploitable resources. If, for example, water needs to be acquired on Mars, it might be extracted from the air, from surface materials containing chemically bound water, or from subsurface ice or permafrost. Which reservoir should be tapped depends on trade-offs between various extraction technologies available and detailed knowledge of the martian environment. The atmospheric abundance of water is known adequately for this purpose, but the location (particularly the depth) of subsurface ice is not.

If there is a requirement to mine water at the landing site, then precursor flights should be designed to locate regions where subsurface ice may exist. Similarly, detailed knowledge of the local mineralogy should be obtained on precursor flights for in situ extraction of water from mined minerals. If habitation is chosen as a long-term goal of Mars exploration, then the technology necessary to locate subsurface water or permafrost will probably need to be developed.

NOTES AND REFERENCES

1. Space Science Board, *A Strategy for Space Biology and Medical Sciences for the 1980s and 1990s*, National Academy Press, Washington, D.C., 1987, Chapter 4.
2. For an assessment of this problem in the context of Space Station Freedom, see Board on Environmental Studies and Toxicology, *Guidelines for Developing Spacecraft Maximum Allowable Concentrations for Space Station Contaminants*, National Academy Press, Washington, D.C., 1992.
3. See Ref. 1, Chapter 2.
4. See Ref. 1, p. 32.
5. Mars Atmosphere Knowledge Requirements Working Group, *SEI Engineering Requirements on Robotic Missions*, Roger D. Bourke (ed.), JPLD-8465, NASA, Jet Propulsion Laboratory, Pasadena, Calif., May 1991.

4

Conclusions

The Committee on Human Exploration finds that a program for the exploration of the Moon and Mars by humans offers both challenges and opportunities for the participation of the scientific community. Foremost is the fact that particular, enabling scientific information is required if a Moon/Mars program is ever to succeed in one of its prime goals, the expansion of human presence and human activity beyond Earth orbit into the solar system. This will remain the case even if a major Moon/Mars program is not initiated for 5 years or 25 years. The information that the committee deems critical is concerned largely with aspects of space biology and medicine and associated characteristics of the radiation environment. This in itself is not a new finding; recognition of the need for such information has been building over the past 30 years with little progress on solutions. What is required is that NASA (and other agencies involved in implementing a human exploration project) make a long-term commitment to sponsoring a rigorous, efficient, high-quality research program on the ground and in space. The resources required will be significant and challenge NASA to structure, market, implement, and ultimately manage an adequate plan.

To enable long-duration human flight to, and operations on, the Moon and Mars, we must obtain critical relevant data. However, we must also consider ab initio that the enabling research has a purpose above and beyond the simplistic, but prime, goal of achieving human presence and implied elementary survival. If a Moon/Mars program is to accomplish more than merely establishing a human presence in space, then achieving the program's yet-to-be-established specific goals and objectives demands that human performance and "pre-presence" preparation be optimized. This imperative places additional weight on the acquisition of scientific data on, for example, the distribution of potential lunar resources, details of the atmosphere of Mars, and information on the physical, chemical, and biological properties of the martian surface.

Science permeates all aspects of human exploration, no matter which architecture is finally selected and regardless of which set of candidate goals and objectives evolves. The involvement of the scientific community is needed to help set the goals for purely robotic missions, to analyze both scientific and engineering data, to structure appropriate tasks for humans, and to assist in the optimal integration of human and robotic activities. This pervasive requirement for scientific input mandates that the piloted spaceflight community develop a new understanding of and attention to the conduct of space science. It simultaneously requires that the scientific community interact constructively with those charged with implementation of a Moon/Mars program. In fact, success will require a technical and programmatic approach that eliminates the historical dichotomy between the "manned" and "unmanned" spaceflight programs.

Bibliography

Advisory Committee on the Future of the U.S. Space Program, *Report of the Advisory Committee on the Future of the U.S. Space Program*, U.S. Government Printing Office, Washington, D.C., 1990.

Committee on Human Exploration of Space, *Human Exploration of Space: A Review of NASA's 90-Day Study and Alternatives*, National Academy Press, Washington, D.C., 1990.

Committee on Space Policy, *Toward a New Era in Space: Realigning Policies to New Realities*, National Academy Press, Washington, D.C., 1988.

Mars Atmosphere Knowledge Requirements Working Group, *SEI Engineering Requirements on Robotic Missions*, Roger D. Bourke, (ed.), JPLD-8465, NASA, Jet Propulsion Laboratory, Pasadena, Calif., May 1991.

Mars Science Working Group, *A Strategy for the Scientific Exploration of Mars*, JPLD-8211, NASA, Jet Propulsion Laboratory, Pasadena, Calif., 1991.

NASA Advisory Council, *Exploring the Living Universe: A Strategy for Space Life Sciences*, Report of the NASA Life Sciences Strategic Planning Study Committee, NASA, Washington, D.C., 1988.

NASA Advisory Council, *Strategic Considerations for Support of Humans in Space and in Moon/Mars Exploration Missions*, Vol. I & II, Life Sciences Research and Technology Programs, Aerospace Medicine Advisory Committee, NASA, Washington, D.C., 1992.

NASA, *Report of the 90-Day Study on Human Exploration of the Moon and Mars*, NASA, Washington, D.C., 1989.

National Commission on Space, *Pioneering the Space Frontier*, The Report of the National Commission on Space, Bantam Books, New York, 1986.

National Council on Radiation Protection and Measurements, *Guidance on Radiation Received in Space Activities*, NCRP Report No. 98, National Council on Radiation Protection and Measurements, Bethesda, Maryland, 1989.

Office of Exploration, *Leadership and America's Future in Space*, A Report to the Administrator by Dr. Sally K. Ride, August 1987, NASA, Washington, D.C., 1987.

Office of Exploration, *Beyond the Earth's Boundaries: Human Exploration of the Solar System in the 21st Century*, NASA, Washington, D.C., 1988.

BIBLIOGRAPHY 49

Office of Space Science and Applications, *Cardiopulmonary Discipline Science Plan,* Life Sciences Division, NASA, Washington, D.C., 1991.

Office of Space Science and Applications, *Controlled Ecological Life Support Systems (CELSS),* Life Sciences Division, NASA, Washington, D.C., 1991.

Office of Space Science and Applications, *Developmental Biology Discipline Plan,* Life Sciences Division, NASA, Washington, D.C., 1991.

Office of Space Science and Applications, *Human Factors Discipline Science Plan,* Life Sciences Division, NASA, Washington, D.C., 1991.

Office of Space Science and Applications, *Musculoskeletal Discipline Science Plan,* Life Sciences Division, NASA, Washington, D.C., 1991.

Office of Space Science and Applications, *Neuroscience Discipline Science Plan,* Life Sciences Division, NASA, Washington, D.C., 1991.

Office of Space Science and Applications, *Regulatory Physiology Discipline Plan,* Life Sciences Division, NASA, Washington, D.C., 1991.

Office of Space Science and Applications, *Space Biology Plant Program Plan,* Life Sciences Division, NASA, Washington, D.C., 1991.

Office of Space Science and Applications, *Space Radiation Health Program Plan,* Life Sciences Division, NASA, Washington, D.C., 1991.

Office of Space Science and Applications, *Space Life Sciences Strategic Plan,* Life Sciences Division, NASA, Washington, D.C., 1992.

Office of Technology Assessment, *Exploring the Moon and Mars: Choices for the Nation,* OTA-ISC-502, U.S. Government Printing Office, Washington, D.C., 1991.

Space Environment Laboratory, *Solar Radiation Forecasting and Research to Support the Space Exploration Initiative,* NOAA Space Environment Laboratory, Boulder, Colo., 1991.

Space Science Board, *HZE-Particle Effects in Manned Spaceflight,* National Academy of Sciences, Washington, D.C., 1973.

Space Science Board, *Post-Viking Biological Investigations of Mars,* National Academy of Sciences, Washington, D.C., 1977.

Space Science Board, *Recommendations on Quarantine Policy for Mars, Jupiter, Saturn, Uranus, Neptune, and Titan,* National Academy of Sciences, Washington, D.C., 1978.

Space Science Board, *Strategy for Exploration of the Inner Planets: 1977-1987,* National Academy of Sciences, Washington, D.C., 1978.

Space Science Board, *Life Beyond the Earth's Environment: The Biology of Living Organisms in Space,* National Academy of Sciences, Washington, D.C., 1979.

Space Science Board, *Origin and Evolution of Life—Implications for the Planets: A Scientific Strategy for the 1980's,* National Academy of Sciences, Washington, D.C., 1981.

Space Science Board, *A Strategy for Space Biology and Medical Sciences for the 1980s and 1990s,* National Academy Press, Washington, D.C., 1987.

Space Science Board, *Space Science in the Twenty-First Century: Imperatives for the Decades 1995 to 2015—Life Sciences,* National Academy Press, Washington, D.C., 1988.

Space Studies Board, *International Cooperation for Mars Exploration and Sample Return,* National Academy Press, Washington, D.C., 1990.

Space Studies Board, *1990 Update to Strategy for the Exploration of the Inner Planets,* National Academy Press, Washington, D.C., 1990.

Space Studies Board, *The Search for Life's Origins: Progress and Future Directions in Planetary Biology and Chemical Evolution,* National Academy Press, Washington, D.C., 1990.

Space Studies Board, *Assessment of Programs in Space Biology and Medicine 1991,* National Academy Press, Washington, D.C., 1991.

Space Studies Board, *Biological Contamination of Mars: Issues and Recommendations,* National Academy Press, Washington, D.C., 1992.

Synthesis Group, *America at the Threshold,* Report of the Synthesis Group on America's Space Exploration Initiative, U.S. Government Printing Office, Washington, D.C., 1991.

Appendix

COMMITTEE ON SPACE BIOLOGY AND MEDICINE

FRED W. TUREK, Northwestern University, *Chair*
ROBERT M. BERNE, University of Virginia, Charlottesville
PETER DEWS, Harvard Medical School
R.J. MICHAEL FRY, Oak Ridge National Laboratory
FRANCIS (DREW) GAFFNEY, Southwestern Medical Center, Dallas
EDWARD GOETZL, University of California Medical Center,
    San Francisco
ROBERT HELMREICH, University of Texas, Austin
JAMES LACKNER, Brandeis University
BARRY W. PETERSON, Northwestern University
CLINTON T. RUBIN, State University of New York, Stony Brook
ALAN L. SCHILLER, Mt. Sinai Medical Center
TOM SCOTT, University of North Carolina, Chapel Hill
WARREN SINCLAIR, National Council on Radiation Protection and
    Measurements
WILLIAM THOMPSON, North Carolina State University, Raleigh
FRED WILT, University of California, Berkeley

COMMITTEE ON SOLAR AND SPACE PHYSICS

MARCIA NEUGEBAUER, Jet Propulsion Laboratory, *Co-Chair*
THOMAS CRAVENS, University of Kansas
JONATHAN F. ORMES, Goddard Space Flight Center
GEORGE K. PARKS, University of Washington
DOUGLAS M. RABIN, National Optical Astronomy Observatories
DAVID M. RUST, Johns Hopkins University
RAYMOND J. WALKER, University of California, Los Angeles
YUK L. YUNG, California Institute of Technology
RONALD D. ZWICKL, National Oceanic and Atmospheric Administration

COMMITTEE ON SOLAR-TERRESTRIAL RESEARCH

DONALD J. WILLIAMS, Applied Physics Laboratory, *Co-Chair*
ALAN C. CUMMINGS, California Institute of Technology
GORDON EMSLIE, University of Alabama
DAVID C. FRITTS, University of Colorado
ROLANDO R. GARCIA, National Center for Atmospheric Research
MARGARET G. KIVELSON, University of California, Los Angeles
DAVID J. McCOMAS, Los Alamos National Laboratory
EUGENE N. PARKER, University of Chicago
JAMES F. VICKREY, SRI International

NOTE: The National Research Council's Committee on Solar-Terrestrial Research (CSTR) and Committee on Solar and Space Physics (CSSP) meet jointly as a federated committee and report directly to their parent National Research Council boards, the Board on Atmospheric Science and Climate for CSTR and the Space Studies Board for CSSP.

COMMITTEE ON PLANETARY AND LUNAR EXPLORATION

LARRY W. ESPOSITO, University of Colorado, *Chair*
RETA BEEBE, New Mexico State University, Las Cruces
ALAN P. BOSS, Carnegie Institution of Washington
ANITA L. COCHRAN, University of Texas, Austin
PETER J. GIERASCH, Cornell University
WILLIAM S. KURTH, University of Iowa, Iowa City
LUCY-ANN McFADDEN, University of Maryland
CHRISTOPHER P. McKAY, NASA Ames Research Center
DUANE O. MUHLEMAN, California Institute of Technology
NORMAN R. PACE, Indiana University
GRAHAM RYDER, Lunar and Planetary Institute
PAUL D. SPUDIS, Lunar and Planetary Institute
PETER H. STONE, Massachusetts Institute of Technology
GEORGE WETHERILL, Carnegie Institution of Washington
RICHARD W. ZUREK, Jet Propulsion Laboratory

COMMITTEE ON MICROGRAVITY RESEARCH

ROBERT F. SEKERKA, Carnegie Mellon University, *Chairman*
ROBERT A. BROWN, Massachusetts Institute of Technology
FRANKLIN D. LEMKEY, United Technologies Research Center
WILLIAM A. SIRIGNANO, University of California, Irvine
THOMAS A. STEITZ, Howard Hughes Medical Institute

Scientific Opportunities in the Human Exploration of Space

Scientific Opportunities in the Human Exploration of Space

Committee on Human Exploration

Space Studies Board

Commission on Physical Sciences, Mathematics, and Applications

National Research Council

NATIONAL ACADEMY PRESS
Washington, D.C. 1994

NOTICE: The project that is the subject of this report was approved by the Governing Board of the National Research Council, whose members are drawn from the councils of the National Academy of Sciences, the National Academy of Engineering, and the Institute of Medicine. The members of the committee responsible for the report were chosen for their special competences and with regard for appropriate balance.

This report has been reviewed by a group other than the authors according to procedures approved by a Report Review Committee consisting of members of the National Academy of Sciences, the National Academy of Engineering, and the Institute of Medicine.

The National Academy of Sciences is a private, nonprofit, self-perpetuating society of distinguished scholars engaged in scientific and engineering research, dedicated to the furtherance of science and technology and to their use for the general welfare. Upon the authority of the charter granted to it by the Congress in 1863, the Academy has a mandate that requires it to advise the federal government on scientific and technical matters. Dr. Bruce M. Alberts is president of the National Academy of Sciences.

The National Academy of Engineering was established in 1964, under the charter of the National Academy of Sciences, as a parallel organization of outstanding engineers. It is autonomous in its administration and in the selection of its members, sharing with the National Academy of Sciences the responsibility for advising the federal government. The National Academy of Engineering also sponsors engineering programs aimed at meeting national needs, encourages education and research, and recognizes the superior achievements of engineers. Dr. Robert M. White is president of the National Academy of Engineering.

The Institute of Medicine was established in 1970 by the National Academy of Sciences to secure the services of eminent members of appropriate professions in the examination of policy matters pertaining to the health of the public. The Institute acts under the responsibility given to the National Academy of Sciences by its congressional charter to be an adviser to the federal government and, upon its own initiative, to identify issues of medical care, research, and education. Dr. Kenneth I. Shine is President of the Institute of Medicine.

The National Research Council was organized by the National Academy of Sciences in 1916 to associate the broad community of science and technology with the Academy's purposes of furthering knowledge and advising the federal government. Functioning in accordance with general policies determined by the Academy, the Council has become the principal operating agency of both the National Academy of Sciences and the National Academy of Engineering in providing services to the government, the public, and the scientific and engineering communities. The Council is administered jointly by both Academies and the Institute of Medicine. Dr. Bruce M. Alberts and Dr. Robert M. White are chairman and vice chairman, respectively, of the National Research Council.

Support for this project was provided by Contract NASW 4627 between the National Academy of Sciences and the National Aeronautics and Space Administration.

Cover: Mars mosaic image courtesy of Alfred McEwen of the U.S. Geological Survey, Flagstaff, Arizona. Lunar crescent image courtesy of Dennis di Cicco. Cover design by Penny Margolskee.

Copies of this report are available from Space Studies Board, National Research Council, 2101 Constitution Avenue, N.W., Washington, D.C. 20418.

Copyright 1994 by the National Academy of Sciences. All rights reserved.

Printed in the United States of America

COMMITTEE ON HUMAN EXPLORATION

NOEL W. HINNERS, Martin Marietta Astronautics Company, *Chair*
RICHARD L. GARWIN,* IBM T.J. Watson Research Center
LOUIS J. LANZEROTTI, AT&T Bell Laboratories
ELLIOTT C. LEVINTHAL,* Stanford University
WILLIAM J. MERRELL, JR., Texas A&M University
ROBERT H. MOSER, University of New Mexico
JOHN E. NAUGLE,[†] National Aeronautics and Space Administration (retired)
GEORGE DRIVER NELSON, University of Washington
SALLY K. RIDE,* University of California, San Diego
MARCIA S. SMITH,[†] Congressional Research Service
GERALD J. WASSERBURG,[†] California Institute of Technology

Staff

DAVID H. SMITH, Executive Secretary
BOYCE N. AGNEW, Administrative Assistant

---

*Former committee member who participated in writing this report.
[†]Committee members added for third CHEX study who participated in writing this report.

SPACE STUDIES BOARD

LOUIS J. LANZEROTTI, AT&T Bell Laboratories, *Chair*
JOSEPH A. BURNS, Cornell University
ANDREA K. DUPREE,* Harvard-Smithsonian Center for Astrophysics
JOHN A. DUTTON, Pennsylvania State University
ANTHONY W. ENGLAND, University of Michigan
LARRY ESPOSITO,* University of Colorado
JAMES P. FERRIS, Rensselaer Polytechnic Institute
HERBERT FRIEDMAN, Naval Research Laboratory
RICHARD GARWIN,* IBM T.J. Watson Research Center
RICCARDO GIACCONI,* IBM T.J. Watson Research Center
HAROLD J. GUY, University of California, San Diego
NOEL W. HINNERS, Martin Marietta Astronautics Company
JAMES R. HOUCK,* Cornell University
DAVID A. LANDGREBE,* Purdue University
ROBERT A. LAUDISE, AT&T Bell Laboratories
RICHARD S. LINDZEN, Massachusetts Institute of Technology
JOHN H. McELROY, University of Texas, Arlington
WILLIAM J. MERRELL, JR., Texas A&M University
RICHARD K. MOORE,* University of Kansas
ROBERT H. MOSER,* University of New Mexico
NORMAN F. NESS, University of Delaware
MARCIA NEUGEBAUER, Jet Propulsion Laboratory
SIMON OSTRACH, Case Western Reserve University
JEREMIAH P. OSTRIKER, Princeton University
CARLE M. PIETERS, Brown University
JUDITH PIPHER, University of Rochester
MARK SETTLE,* ARCO Oil Company
WILLIAM A. SIRIGNANO, University of California, Irvine
JOHN W. TOWNSEND, National Aeronautics and Space Administration (retired)
FRED W. TUREK, Northwestern University
ARTHUR B.C. WALKER, JR., Stanford University

MARC S. ALLEN, Director

---

*Former member.

COMMISSION ON PHYSICAL SCIENCES, MATHEMATICS, AND APPLICATIONS

RICHARD N. ZARE, Stanford University, *Chair*
RICHARD S. NICHOLSON, American Association for the Advancement of Science, *Vice Chair*
STEPHEN L. ADLER, Institute for Advanced Study
JOHN A. ARMSTRONG, IBM Corporation (retired)
SYLVIA T. CEYER, Massachusetts Institute of Technology
AVNER FRIEDMAN, University of Minnesota
SUSAN L. GRAHAM, University of California, Berkeley
ROBERT J. HERMANN, United Technologies Corporation
HANS MARK, University of Texas, Austin
CLAIRE E. MAX, Lawrence Livermore National Laboratory
CHRISTOPHER F. McKEE, University of California, Berkeley
JAMES W. MITCHELL, AT&T Bell Laboratories
JEROME SACKS, National Institute of Statistical Sciences
A. RICHARD SEEBASS III, University of Colorado
CHARLES P. SLICHTER, University of Illinois, Urbana-Champaign
ALVIN W. TRIVELPIECE, Oak Ridge National Laboratory

NORMAN METZGER, Executive Director

Preface

In 1988 the National Academy of Sciences and the National Academy of Engineering stated in the report, *Toward a New Era in Space: Realigning Policies to New Realities,* that "the ultimate decision to undertake further voyages of human exploration and to begin the process of expanding human activities into the solar system must be based on nontechnical factors." It is clear, however, that if and when a program of human exploration is initiated, the U.S. research community must play a central role by providing the scientific advice necessary to help make the relevant political and technical decisions.

Since its establishment in 1958, the Space Studies Board (SSB; formerly the Space Science Board) has been the principal nongovernmental advisory body on civil space research in the United States. In this capacity, the board established the Committee on Human Exploration (CHEX) in 1989 to examine many of the science and science policy matters concerned with the return of astronauts to the Moon and eventual voyages to Mars. The board asked CHEX to consider three major questions:

1. What scientific knowledge must be obtained as a prerequisite for prolonged human space missions?
2. What scientific opportunities might derive from prolonged human space missions?
3. What basic principles should guide the management of both the prerequisite science activities necessary to enable human exploration and

the scientific activities that may be carried out in conjunction with human exploration?

This report focuses on the second of these topics. The first topic was covered in *Scientific Prerequisites for the Human Exploration of Space,* published in 1993; the third topic is the subject of a future report.

The Space Studies Board and CHEX concluded that the existing research strategies of several of the board's discipline committees form a basis for *beginning* to determine the scientific research opportunities that might arise if and when humans undertake voyages to the Moon and Mars. (See the appendix for a list of these committees and their contributing members.) CHEX thus asked the discipline committees to identify those scientific opportunities and classify them under two headings: (1) those that can be conducted only in association with long-term human missions and (2) those that could also be conducted by other means (for example, robotic or ground-based) to achieve the same or equivalent goals.

Early in their analyses the discipline committees found that, with one exception, they were not able to identify opportunities that unambiguously *require* human presence. The exception, the study of the effects of prolonged missions to the Moon and Mars on human physiology and psychology, is in and of itself of low priority absent a program of human exploration. Regarding opportunities that are in competition with other means, difficulty was encountered because of the considerable uncertainty existing concerning the practical capability of humans and the eventual capabilities of robotic missions over the long time scale involved in any program of human exploration. The committees thus expanded their advice to include the following considerations:

1. Identification of those scientific objectives for the Moon and Mars for which human presence can play a significant role;
2. Discussion of the realistic capabilities of humans and robots in planetary exploration and in carrying out scientific investigations in those environments;
3. Discussion of the appropriate phasing and mix of human and robotic activities in achieving those objectives;
4. Discussion of the requirements for crew selection and training, technical development, and program structure to meet the scientific objectives in a program of human exploration; and
5. Identification of robotic scientific opportunities that may be enabled by some of the technology developed for the human exploration program.

CHEX itself developed a description of the overall role of science in a program of human exploration. In that context, it then assimilated, evaluated, and integrated the contributions of the discipline committees. Information on the biomedical research opportunities arising from prolonged space missions was provided by the SSB's Committee on Space Biology and Medicine. Input on field science, the relative capabilities of humans and robots, and the search for planets around other stars was supplied by the SSB's Committee on Planetary and Lunar Exploration. (CHEX consulted *A Strategy for the Scientific Exploration of Mars,* by the National Aeronautics and Space Administration's Mars Science Working Group, for additional information on the planetological and exobiological aspects of Mars precursor science.) Research opportunities in astrophysics and solar and space physics were considered by the SSB's Committee on Solar and Space Physics and the Board on Atmospheric Sciences and Climate's Committee on Solar-Terrestrial Research. Astronomical input from these discipline committees was augmented with material from *The Decade of Discovery in Astronomy and Astrophysics*, a report written by the National Research Council's Astronomy and Astrophysics Survey Committee. Details of the individual scientific strategies and goals of the relevant discipline committees, on which they based much of their input, are contained in the reports listed in the bibliography.

    Noel W. Hinners, *Chair*
    Committee on Human
    Exploration

Contents

EXECUTIVE SUMMARY 1

1 SPACE SCIENCE AND HUMAN EXPLORATION OF
THE SOLAR SYSTEM 3
Enabling Science, 5
Enabled Science, 6
References, 7

2 ROBOTS AND HUMANS: AN INTEGRATED APPROACH 9
Relative Advantages, 9
Relative Limitations, 10
The Optimal Mix of Humans and Robots, 11
Science Precursor Missions, 12
Technology to Optimize the Scientific Return, 14
References, 15

3 SCIENCE ENABLED BY HUMAN EXPLORATION 17
Field Science, 17
Unraveling Solar Particle Emission History, 19
The Search for Life on Mars, 20
Impact History of the Terrestrial Planets, 21
Martian Climate History, 22

Emplacement and Attendance of Large or Complex
   Instruments, 22
   Detection and Study of Other Solar Systems, 24
   Study of High-Energy Cosmic Rays, 25
   Advanced Pinhole Occulter, 25
Life Sciences, 26
Science Enabled by Technology Developed for a
   Moon/Mars Program, 26
Scientific Community Participation, 27
References, 28

BIBLIOGRAPHY   30

APPENDIX: PARTICIPATING DISCIPLINE COMMITTEES   33

Executive Summary

What role should the scientific community play if a political decision is made to initiate a program for the human exploration of the Moon and Mars? As the first phase of its study to answer this question, the Committee on Human Exploration (CHEX) found that certain critical scientific information is needed before humans can safely return to the Moon for extended periods and, eventually, undertake voyages to Mars.[1] In addition to the scientific challenges of ensuring human survival in space, CHEX found that a Moon/Mars program offers "opportunities for the participation of the scientific community."[2] What are these opportunities? What, if any, scientific research is "enabled" by the existence of a program of human exploration of the Moon and Mars? Does the technology developed for a Moon/Mars program open new avenues for scientific research?

In attempting to answer these questions, CHEX reached the following conclusions:

1. Given that a program of human exploration is undertaken primarily for reasons other than scientific research, humans can make significant contributions to scientific activities through their ability to conduct scientific field work and by using their capabilities to emplace and attend scientific facilities on planetary bodies.

2. The fractional gravity environment of the Moon and Mars and of space vehicles in transit to and from Mars offers a unique opportunity to study the effects of prolonged exposure to fractional gravity levels on living

systems. Similarly, space missions lasting as long as 2 to 3 years will provide an unusual opportunity to study human behavior under uniquely stressful conditions (confinement with no immediate possibility of escape). The committee emphasizes, however, that both of these possibilities are at this time not inherently of high scientific priority in the absence of a program of human exploration.

3. There will be significant limitations on humans performing scientific activities because of safety concerns and the restrictions on mobility and manipulation imposed by the design of current spacesuits. Technology development is required to improve spacesuits, biomedical diagnostic procedures, life support systems (both open and closed), and tools.

4. With the robotic technology expected to be utilized over the next few decades, using robots to perform certain scientific activities (e.g., field work) on extraterrestrial planetary surfaces will not be a realistic alternative to having humans on site. Technology development is required to improve both the capability of robotic field aids and the ability to control them remotely.

5. The next steps in the exploration of Mars should be carried out by robotic spacecraft controlled from Earth. As the program evolves to include human exploration, the optimal mix of human and robotic activities is likely to include proximate human control of robots with a shorter time delay than can be achieved from Earth.

6. Space scientists in non-planetary science disciplines will be in the best position to take advantage of the scientific opportunities enabled by a Moon/Mars program if there is a steady, phased program of scientific projects on Earth and in Earth orbit.

7. Astronauts with a high level of relevant scientific knowledge and experience must be included in Moon/Mars missions. Crew training and exploration planning should be designed to take advantage of human initiative, flexibility, adaptability, and deductive and inductive reasoning abilities.

8. Scientists must be involved in every stage of a Moon/Mars program from conception to execution to ensure that quality science is accomplished, the science supported best takes advantage of human presence, and resources available to the whole of space science are competitively allocated.

REFERENCES

1. Space Studies Board, *Scientific Prerequisites for the Human Exploration of Space*, National Academy Press, Washington, D.C., 1993.
2. Space Studies Board, *Scientific Prerequisites for the Human Exploration of Space*, National Academy Press, Washington, D.C., 1993, page 46.

1
Space Science and Human Exploration of the Solar System

The post-Apollo directions of a U.S. program of human exploration of the solar system have long been the subject of study, discussion, debate, and controversy. Most concepts for the next steps in the human exploration of space, including those going back as far as the mid-1960s, have focused on missions to the Moon and Mars and their immediate vicinity.[1-6] Those studies were conducted largely in the context of a future program of human exploration of the Moon and Mars that was assumed to be inevitable.

Political support, however, has not materialized for initiating a piloted return to the Moon or for journeying to Mars; in fact, it has been difficult to get a political consensus to support the funding of a space station, the prime goal of which is, arguably, to prepare for long-duration human space exploration. The arguments, pro and con, for continued human spaceflight have shifted as the basic rationale has changed. No longer is competition with the Soviet Union a compelling force as it was for Apollo, and the economic pressures faced by the nation are causing many to question whether this is the time for human exploration of the solar system.

Despite the current uncertainty, however, the possibility for future human exploration of the Moon and Mars remains. In this regard, the Committee on Human Exploration (CHEX) recognizes that political factors can change rapidly and can have profound effects on the pace and content of a human space exploration program, as they did when President Kennedy committed to the Apollo program. CHEX views the current interlude as an opportune time in which to calmly and methodically study and stipulate the role of science in any future program of human exploration of the solar system.

Given the often lofty, but still ill-defined, human exploration aspirations, what is the role of science in a Moon/Mars program? CHEX started with the recognition that one of the major goals in the National Aeronautics and Space Administration's (NASA) original charter was the acquisition of new scientific knowledge about space and the terrestrial environment. Indeed, scientific goals have always played an important part in NASA's activities. Thus it is natural to expect that science will play a major role in any future program of human exploration, as it did in Apollo and in all subsequent piloted spaceflight programs. The specific nature of that role and the way in which the scientific community has historically interacted with human space exploration will be dealt with in the third CHEX report.

It is not surprising then, that many, if not all, concepts for human exploration of the Moon and Mars include scientific investigations. Many proponents also propose using the Moon as an observational platform from which to conduct astronomical and space physics studies.

Is science then *the* motivation for a Moon/Mars program? This question was answered in the negative by the National Academy of Sciences and National Academy of Engineering in a report on space policy prepared in 1988. It stated that "the ultimate decision to undertake further voyages of human exploration and to begin the process of expanding human activities into the solar system must be based on nontechnical factors."[7] In other words, the expansion of human presence and activity into the solar system does not demand any a priori scientific research component beyond the enabling research needed to provide for the health and safety of the astronauts (see next section).

Nevertheless, recognizing the need for enabling research and that piloted flight can result in new or modified space science opportunities, the U.S. research community has the opportunity and obligation to provide the best and most constructive scientific advice it can to help shape the political and technical decisions regarding piloted flight.[8] Accepting such a role commits scientists to participating in establishing human exploration strategy and goals, mission planning, management, implementation, and analysis of results. During mission design and operations, scientists must participate to ensure optimal scientific return. Part of that optimization is the inclusion in the crews not only of people trained to perform particular scientific tasks, but also of experienced scientists. Indeed, scientist-astronauts have an important role to play in planning, postmission analysis, and preparations for future exploration.

Since the end of the Apollo program in 1972, humans have not set foot on another body in the solar system. The Apollo experience involved hundreds of scientists in many disciplines. Although science was not emphasized or well planned at the beginning of the program, Apollo evolved a highly successful mechanism to include scientific input that ultimately produced important scientific results.

Participation of scientists in a program of human exploration is a sensitive subject in the broad scientific community. Some individuals fear that *any* involvement is an implicit endorsement of such a program. Others fear that science is or will be used as a justification or that low-priority and/or low-quality science will be funded under the umbrella of an expensive human spaceflight program. Indeed, experience shows that these concerns cannot be dismissed out of hand—thus part of the Space Studies Board's (SSB) rationale for establishing CHEX was to ensure that the scientific aspects of a Moon/Mars program are established in the proper context. That is, only science that truly takes unique advantage of human presence should be undertaken and then only if it is of competitive quality.

ENABLING SCIENCE

CHEX concluded in its first study that the most important responsibility facing the scientific community, in the initial stages of a program of human exploration, is to define the conditions necessary to maintain the health and safety and ensure the optimal performance of astronauts during exploration missions. Answers are urgently needed to such questions as, Can humans function effectively on the Moon for long periods? and, Can they survive the lengthy journey to Mars? CHEX identified these *enabling* science issues in its first report[9] and classified them according to their degree of urgency.

Critical research issues were defined as those for which inadequate scientific data lead to unacceptably high risks to any program of extended space exploration by humans. They are the potential "showstoppers" for a Moon/Mars project. Items in this first category include, for example, the effects of prolonged exposure of humans to the microgravity and space radiation environments.

Optimal performance issues, the second category, were defined as those that, based on current knowledge, do not appear to pose serious dangers to the health and well-being of humans in space. They could, however, reduce human performance in flight or on planetary surfaces and result in a less than optimal return from the mission. Research to understand these factors cannot be neglected, and some of them may become critical research issues relative to long-duration human spaceflight and return to terrestrial gravity, when extraterrestrial habitation is considered or when new research information is obtained.

ENABLED SCIENCE

Given an eventual political decision to undertake a Moon/Mars program, how might prolonged human space voyages *enable* or enhance the accomplishment of overall space science objectives? Before addressing this question, CHEX reiterates the earlier position of the Space Studies Board that a program of solar system exploration that includes only the Moon and Mars and their immediate vicinity is scientifically incomplete.[10] The obvious concern is that a program of human exploration, which by its very nature would be expensive, could dominate NASA budgetarily, managerially, and programmatically to the detriment of a balanced scientific program. The existence of a vigorous ongoing space science program can go a long way toward creating a receptive environment for a program of human exploration.

Many of the scientific objectives for the Moon and Mars are a subset of the general goals for the scientific exploration of the solar system outlined in past SSB reports. For the Moon and Mars in general, we seek to learn their thermal, magmatic, and tectonic evolution; their bombardment history; and the origin and evolution of their volatiles. We hope to learn about the origin of the Moon and its relationship to the Earth. For Mars we strive to understand the history of its climate, the processes of surface weathering and modification, and global aspects of the magnetic field and associated interactions with the interplanetary medium. Also of high priority is understanding the history of martian biogenic elements and determining whether life ever existed there.[11–13]

Human exploration of the Moon and Mars might also lead to the achievement of objectives in fields other than the planetary sciences. Studies of the lunar regolith and martian ice cores may, for example, reveal the long-term evolution of the particle and photon outputs of the Sun. Similarly, if a human exploration program includes the construction and operation of scientific observatories on the Moon, it might, for example, aid our understanding of the mechanisms operating in solar flares, the origin of very high energy cosmic rays, and the frequency of occurrence of planets around other stars.[14–16]

A Moon/Mars program might enable studies of the response of living organisms to microgravity and fractional gravity environments.[17] In addition, crews on Mars exploration missions will experience a combination of circumstances, including prolonged sequestration with no immediate possibility of escape, that might enable unique studies of human behavior.[18] It must be stressed, however, that these research opportunities in the life sciences are fundamentally different from those in the physical sciences (outlined above), because the latter are inherently of high scientific priority to their relevant research communities, whereas the former are currently not, absent a program of human exploration.

Over the years the SSB has made many specific recommendations for scientific investigations in space, but none of the board's previous reports considered possible opportunities in the physical or biomedical sciences enabled by prolonged human space missions. For this report, CHEX considered ways in which human presence might enhance the accomplishment of previously recommended robotic scientific investigations and also considered what new investigations, consistent with the SSB's scientific strategies, might be enabled by a human exploration program. From this extended list, CHEX selected a number of specific examples that have valid scientific and technical reasons for being performed in conjunction with a Moon/Mars program and that would

- Be enhanced or enabled by prolonged human space missions, and
- Contribute in a major way to achieving the overall goals of space science.

The investigations in the physical sciences described in Chapter 3 meet these criteria. But, as is discussed below, some of those suggested in other reports do not. This observation raises a major concern of the scientific community—too often little or no competitive analysis and prioritization have been done, with respect to alternative modes or other science, to assess the merit of the proposed science for a Moon/Mars program.[19]

REFERENCES

1. President's Science Advisory Committee, Joint Space Panels, *The Space Program in the Post-Apollo Period,* U.S. Government Printing Office, Washington, D.C., February 1967.
2. NASA, *Beyond the Earth's Boundaries: Human Exploration of the Solar System in the 21st Century,* NASA, Washington, D.C., 1988.
3. Advisory Committee on the Future of the U.S. Space Program, *Report of the Advisory Committee on the Future of the U.S. Space Program* (the "Augustine report"), U.S. Government Printing Office, Washington, D.C., 1990.
4. NASA, *Leadership and America's Future in Space,* NASA, Washington, D.C., 1987.
5. Synthesis Group, *America at the Threshold,* Report of the Synthesis Group on America's Space Exploration Initiative, U.S. Government Printing Office, Washington, D.C., 1991.
6. NASA, *Report of the 90-day Study on Human Exploration of the Moon and Mars,* NASA, Washington, D.C., 1989.
7. Committee on Space Policy, *Toward a New Era in Space: Realigning Policies to New Realities* (the "Stever report"), National Academy Press, Washington, D.C., 1988.
8. Space Studies Board, *Scientific Prerequisites for the Human Exploration of Space,* National Academy Press, Washington, D.C., 1993, page 2.
9. Space Studies Board, *Scientific Prerequisites for the Human Exploration of Space,* National Academy Press, Washington, D.C., 1993, pages 3-4.

10. Space Studies Board, *1990 Update to Strategy for the Exploration of the Inner Planets,* National Academy Press, Washington, D.C., 1990.

11. Space Studies Board, *Strategy for Exploration of the Inner Planets: 1977-1987,* National Academy of Sciences, Washington, D.C., 1978.

12. Space Studies Board, *1990 Update to Strategy for the Exploration of the Inner Planets,* National Academy Press, Washington, D.C., 1990, Chapters 5 and 6.

13. Space Studies Board, *The Search for Life's Origins: Progress and Future Directions in Planetary Biology and Chemical Evolution,* National Academy Press, Washington, D.C., 1990.

14. Astronomy and Astrophysics Survey Committee, *The Decade of Discovery in Astronomy and Astrophysics,* National Academy Press, Washington, D.C., 1991.

15. Space Studies Board, *Assessment of Programs in Solar and Space Physics 1991,* National Academy Press, Washington, D.C., 1991.

16. European Space Agency, *Mission to the Moon: Europe's Priorities for the Scientific Exploration and Utilization of the Moon,* Report of the Lunar Study Steering Group, ESA SP-1150, European Space Agency, Noordwijk, The Netherlands, June 1992.

17. Space Studies Board, *Assessment of Programs in Space Biology and Medicine 1991,* National Academy Press, Washington, D.C., 1991.

18. Space Studies Board, *Assessment of Programs in Space Biology and Medicine 1991,* National Academy Press, Washington, D.C., 1991, Chapter 4.

19. See, for example, Nancy Ann Budden and Paul D. Spudis, "SEI Science: Measuring the Return," *Aerospace America,* March 1993, page 22.

2

Robots and Humans: An Integrated Approach

Most concepts for Moon/Mars exploration envision a mix of robots and humans. However, the criteria for deciding how each of them should be used, and in what combination, are not usually stated and probably were never formally developed. The result is that the concepts are biased according to the background of the study group; human exploration advocates tend to minimize the use of robots, whereas traditional space scientists tend to downplay the potential of human presence. CHEX believes that decisions regarding the mix of robots and humans to explore the Moon and Mars, and to carry out other scientific investigations in space, should be made with explicit cognizance of the relative strengths and weaknesses of each evaluated in the context of well-defined and specific tasks to be performed.

RELATIVE ADVANTAGES

Human presence can bring to planetary exploration a level of capability representing an essential aspect of scientific methodology: an iterative process of observing, hypothesizing, testing, and synthesizing. Activities ideally suited to humans include those requiring the techniques of intensive field study and tasks requiring complex, physical articulation combined with expert knowledge and the ability to adapt to new situations. Humans conducting scientific observations on planetary surfaces can perform their work with an inherent flexibility not easily equaled by the more cumbersome and delay-ridden methods of remote control, especially at significant radio-delay distances (for example, at Mars). Assessment of complex natural systems makes excellent use of the human capability for serendipitous discovery and response. This human advantage is, for the time being, taken to pertain also to the activities of machines manipulated remotely by humans in near-real-time (that is, in a relatively local control loop with a short time delay).

Robots have several obvious advantages. They are inherently expendable and thus should be used in situations in which the risk to humans is excessive or for which there is no clear advantage to using humans. Robots excel at performing repetitive, tedious tasks that are amenable to programming and that do not need or take advantage of unique human capabilities. Lastly, robots can have a duty cycle that is uninterrupted by the need to rest, sleep, or perform the mundane tasks that devour so much time in the everyday life of humans.

RELATIVE LIMITATIONS

Although humans offer specific advantages in the exploration of planetary surfaces, they have their limitations as well. Because of the harsh environments of the Moon and Mars and the amount of challenging physical work involved, safety considerations will always constrain the amount of time available for people to explore and perform scientific tasks. Humans working in spacesuits will always have less mobility and flexibility than humans working on Earth, despite anticipated improvements in spacesuits. In addition, scientific activities are not the only things people will be doing during human exploration missions. Routine maintenance of the habitat and other equipment is likely to occupy a significant fraction of the astronauts' time (as has become apparent for space station activities). Because of the broad range of scientific investigations proposed for human exploration, the crew (like robots) will not be expert in all relevant activities, although every attempt should be made to select crews that are highly qualified scientifically. Lastly, as was demonstrated in the Chernobyl nuclear accident, the potential for rapid human reaction in response to a local stimulus or observation has a concomitant potential for rapidly introducing errors.

Robots likewise have limitations. The creation of nearly autonomous machines with humanlike cognitive abilities continues to elude the robotic research community and may well do so for a considerable time into the future. At the moment, robots are capable of only simple manipulation; techniques for human-quality dexterity have yet to be demonstrated. Given current capabilities, robots require considerable human control and interaction to accomplish most scientific tasks. Their capabilities are appropriate for simple reconnaissance or prescribed activities in which no major difficulties are encountered. Whether their capabilities will remain at this level will depend on advances in robotic technology prior to the initiation of a program of human exploration. Lastly, even though robots are inherently expendable relative to humans, their cost can be sufficiently large that they ought not be exposed to excessive risk. This limitation can be overcome to the degree that inexpensive robots are developed.

THE OPTIMAL MIX OF HUMANS AND ROBOTS

As a result of its deliberations, CHEX is convinced that the humans-versus-robots controversy is outmoded. The space program has perpetuated this antiquated either/or dichotomy for too long. Examining various aspects of exploration in terrestrial situations clearly shows the proper approach to be a mix.

Considerable experience has been gained in assessing the relative capabilities of humans and robots operating in hostile environments for the location, development, and operation of underwater oil and gas fields. Divers are used primarily to perform tasks beyond the manipulative capability of robots. Robots are used, increasingly, to perform programmable repair tasks and to assess the physical state of systems. Similarly, robots are increasingly used in the hazardous environments presented by nuclear accidents and hazardous waste cleanup. Clearly, safety and risk minimization are paramount determinants in terrestrial situations; no less should be acceptable in human space exploration.

A particularly germane example of the mix of human and robotic activities is in undersea exploration. Even though their exact role is still actively debated,[1] robots are routinely used in oceanographic surveys to scan the ocean floor, emplace sensors, and collect samples. Even when human presence is desired, scientists do not usually study the deep ocean bottom in diving suits (read "spacesuits") but, rather, in pressurized submersibles using teleoperated manipulators and/or robotic devices to probe and acquire samples. The analogy to potential lunar and martian exploration by humans and robots is clear: a synergistic mix based on safety, efficiency, and cost-effectiveness must be the goal.

Given the relative strengths and weaknesses of humans and robots, CHEX envisages that their relative roles in a Moon/Mars program will evolve as knowledge increases and as technological capabilities advance. The initial phases, largely an extension of current space science and involving such activities as global orbital reconnaissance and the deployment of geophysical and meteorological networks, will be conducted exclusively by robots controlled from Earth or operating with varying degrees of autonomy. Further technical developments are needed in both robotics and operational capabilities (e.g., life support systems and exploration tools) to permit humans to survive and function effectively on planetary surfaces. These will lead to a subsequent phase consisting first of a mix of advanced robotic missions, such as those designed to return samples from Mars to Earth for analysis, and, eventually, the first human expeditions.

CHEX envisions further evolution into advanced exploration performed by a synergistic mix of humans and sophisticated robots. Such a mix could, for example, include human operation on Mars supported by robots teleoperated in near-real-time by astronauts on, or in orbit around, Mars.

One might think that an important issue bearing on the relative contributions of humans and robots in a Moon/Mars program would be cost-effectiveness. Ideally, the relative mix of humans and robots used for achieving a particular scientific goal would be based on cost-effectiveness. The concept of cost-effectiveness is, however, difficult to adhere to in a human exploration program, because even though it is axiomatic that robotic missions would cost less than those involving humans, the basic decision to proceed with human exploration is not rooted in science. In that light, CHEX recognizes that at any given time opportunity plays a significant role in prioritizing scientific projects and selecting means of implementation.

Rather than dwell on cost-effectiveness, a more realistic principle, stated in the first CHEX report, is that, "Robotic options should be used until they provide enough information to . . . define a set of scientifically important tasks that can be *well* performed by humans in situ. . . . It cannot be demanded that these tasks be best and most cost-effectively performed by humans."[2] Subsequently, a mix of robots and humans should be used to optimize performance from both a scientific and a safety point of view.

SCIENCE PRECURSOR MISSIONS

Much information about the Moon and Mars has been collected by the Ranger, Surveyor, Lunar Orbiter, Luna, Apollo, Mariner, and Viking missions. However, an orderly series of future robotic missions will be required for collection of data relevant to human safety, for site selection, and for the effective identification and development of enabled scientific opportunities. Such a series of robotic missions would include many that would be a normal complement of an ongoing robotic planetary science program.

For the Moon, several robotic missions are desirable, especially for site selection. A high-resolution global chemical and mineralogical survey of the Moon will allow a much more complete understanding of the variety of lunar geologic features, their origin, and their evolution. Such a survey will also allow for extrapolation of Apollo and Luna data and is needed for targeting more detailed local investigation. Robotic sample returns will greatly aid in further refining site selection and planning scientific investigations. Moreover, a global geophysical network, deployed by landers, will greatly increase our ability to weave the characteristics of the interior into an understanding of the surface evolution and the origin of the Moon.[3]

The pioneering observations performed by the Mariner and Viking missions to Mars were to have been extended by Mars Observer. This remote sensing orbiter mission was designed to characterize martian global geochemistry and the general circulation of the atmosphere. Its high-resolution imaging capabilities, important for geological studies, would also have been useful for selecting future landing sites and planning surface operations. The failure of Mars Observer in August 1993 is therefore a major setback to the scientific exploration of Mars, and the accomplishment of its objectives remains a high scientific priority.

Assuming that a recovery program leads to the accomplishment of some or all of the Mars Observer objectives, a next step in the robotic exploration of Mars should be in situ robotic investigations of its geophysical and meteorological properties. Seismic activity should be explored for its intrinsic scientific value and to define more refined experiments that humans would emplace. Meteorological measurements are required to characterize the atmospheric boundary layer through which the key exchanges of energy, volatiles, and dust occur. The Viking landers made measurements at only two sites and had no capability to measure such important properties as water vapor concentration or to follow up on the discovery of chemical reactivity of the surface material.[4]

To take best advantage of human capabilities in scientific exploration, it will be desirable, some argue essential, to return reconnaissance samples from Mars prior to human exploration. Such sample return missions must deal with the obvious issues associated with planetary quarantine (both forward- and back-contamination).[5] Returned samples will also address potential toxicity issues associated with the highly oxidizing properties of martian soil. This problem may also be tackled by in situ chemical analysis on robotic missions. Possibly more important, precursor sample returns will lead to a major increase in our knowledge of martian processes and history. This will permit a more informed choice of the landing sites for human missions and the types of investigations to be conducted during surface exploration. The Space Studies Board has recommended that "the next major phase of Mars exploration for the United States involve detailed in situ investigations of the surface of Mars and the return to Earth for laboratory analysis of selected martian surface samples."[6]

Stepping-stone missions, or "waypoints" in the language of the Synthesis Group's report, may provide significant scientific return and at the same time help to develop the technological capabilities required to get humans to Mars.[7] For example, possible waypoints are human exploration of a near-Earth asteroid or the martian moons Phobos and Deimos.[8-10] An asteroid mission could be used to test a Mars transfer vehicle and provide useful operational experience in deep space.

TECHNOLOGY TO OPTIMIZE THE SCIENTIFIC RETURN

CHEX recognizes that a program of human exploration would present an opportunity for major advances in our understanding of the Moon and Mars. To realize that potential, high-quality science must be an integral part of the exploration. The optimal strategy for accomplishing the associated science over the next several decades cannot be developed yet because of the uncertain prospects for advances in robotic systems and artificial intelligence.

Major improvements in the human-machine interface of the type needed for the scientific activities discussed below require a focused program dedicated to the challenge of extending human capabilities in hostile environments by developing remote control techniques. A Moon/Mars program cannot rely totally on the development of robotics for terrestrial use. Robotic systems developed, for example, to replace a human welder on an assembly line will not be adequate to function as an extension of humans engaged in field work or maintaining complex instruments on the Moon or Mars. Special features not currently found in industrial robots, such as high-resolution stereoscopic vision and multispectral imaging, would most likely be required to conduct robotically assisted geological field work.[11,12] Coincident with the development of suitable robotics, one must address their effective use. For example, what and how much information should be transmitted to the human operator, and how large a time delay in the human-machine control loop can be tolerated?

The extent to which a human exploration program is able to drive the development of more capable robotic systems over the next several decades, coupled with improved spacesuits (and development of mobile pressurized environments with teleoperations capability enabling humans to perform field work without the encumbrances of a spacesuit), will contribute to determining the optimal mix of humans and machines. Developments in robotics for use in hostile terrestrial environments (deep-sea exploration and activities in "hot" nuclear environments are examples already cited above) will be of great value.

The biomedical research enabled by human exploration will also demand certain technological developments. Prime among these is the need to develop sophisticated, compact diagnostic equipment (some with telemetering capability) to perform essential studies on the responses of the crew and other living organisms to prolonged exposure to the environment of the spacecraft. Such equipment might also serve an important health and safety role in the event of accident or illness in the crew.

The call for technology development could appear obvious and gratuitous; it might be expected that such would occur as a normal consequence of a well-structured plan for both scientific and human exploration. That has not, generally, happened. Study after study, several specifically dealing with the issue,[13,14] has urged greatly increased (by a factor of three) funding and more focused technology development by NASA and a more effective methodology for using existing and future funding. That not much progress has been made can be attributed to a combination of many factors, not all of which are under NASA's control: bureaucratic inertia, organizational conflicts, persistence of irrelevant technologies, low priority relative to near-term flight programs, inadequate justification of the need, lack of an appropriate requirement for an approved program, and political fear of enabling future programs. This combination of somewhat disconnected reasons begs for top-level, determined attention inside and outside of NASA. Without such attention, the committee is pessimistic that the United States will enjoy in the future the leadership in human and robotic space exploration that it has demonstrated in the past.

REFERENCES

1. See, for example, Paul J. Fox and Craig E. Dorman, *"Alvin* and Deep Ocean Research" (letter), *Science,* **261**, July 2, 1993.
2. Space Studies Board, *Scientific Prerequisites for the Human Exploration of Space,* National Academy Press, Washington, D.C., 1993, page 9.
3. Space Studies Board, *1990 Update to Strategy for the Exploration of the Inner Planets,* National Academy Press, Washington, D.C., 1990, pages 18-19.
4. Space Studies Board, *1990 Update to Strategy for the Exploration of the Inner Planets,* National Academy Press, Washington, D.C., 1990, pages 21-24.
5. Space Studies Board, *Biological Contamination of Mars: Issues and Recommendations,* National Academy Press, Washington, D.C., 1992.
6. Space Studies Board, *International Cooperation for Mars Exploration and Sample Return,* National Academy Press, Washington, D.C., 1990, pages 1, 3, and 25.
7. Synthesis Group, *America at the Threshold,* Report of the Synthesis Group on America's Space Exploration Initiative, U.S. Government Printing Office, Washington, D.C., 1991, page A-9.
8. NASA, *Beyond the Earth's Boundaries: Human Exploration of the Solar System in the 21st Century,* NASA, Washington, D.C., 1988, page 32.
9. Synthesis Group, *America at the Threshold,* Report of the Synthesis Group on America's Space Exploration Initiative, U.S. Government Printing Office, Washington, D.C., 1991, page A-37.
10. NASA, *Science Exploration Opportunities for Manned Missions to the Moon, Mars, Phobos, and an Asteroid,* NASA Office of Exploration Doc. No. Z-1.3-001 (also JPL Publication 89-29), NASA, Washington, D.C., 1989.
11. G. Jeffrey Taylor and Paul D. Spudis, "A Teleoperated Robotic Field Geologist," *Engineering, Construction, and Operations in Space II: Proceedings of Space '90,* American Society of Civil Engineers, New York, 1990.
12. Paul D. Spudis and G. Jeffrey Taylor, "The Roles of Humans and Robots as Field

Geologists on the Moon," *Lunar Bases and Space Activities of the 21st Century, 2nd Symposium,* LPI Contribution 652, Lunar and Planetary Institute, Houston, Texas, 1990.

13. Aeronautics and Space Engineering Board, Committee on Advanced Space Technology, *Space Technology to Meet Future Needs,* National Academy Press, Washington, D.C., 1987.

14. Space Studies Board, Aeronautics and Space Engineering Board, Committee on Space Science Technology Planning, *Improving NASA's Technology for Space Science,* National Academy Press, Washington, D.C., 1993.

3

Science Enabled by Human Exploration

Given the scientific goals of space science and the relative capabilities of robots and humans, CHEX has identified two areas in which human presence can enhance important scientific opportunities: (1) field studies of planetary surfaces and (2) the construction and maintenance of large and/or complex scientific instruments. Both of these areas can benefit from human cognitive abilities and from the flexibility provided by in situ or proximate human presence. Additional scientific opportunities arise in the study of the physiological response of living organisms to microgravity and fractional gravity environments and in studies of human behavior during protracted sequestration and other stressful situations. Moreover, technology developed for a human exploration program may enable unrelated robotic space science missions.

FIELD SCIENCE

Field work, a collection of activities in which processes and materials are studied in their natural setting, is intrinsic to several natural sciences, especially geology and biology. Humans bring unique capabilities to field studies: discovery and response accommodate the unexpected and allow the opportunity to redesign an approach. Human presence allows real-time testing of hypotheses using techniques ranging from simple manipulation to conducting a well-designed in situ experiment. Initiative and inductive and deductive thinking are uniquely human capabilities. People innovate and anticipate; their thought processes allow them to distinguish the trivial from the important. Humans are capable of intuitive leaps based on incomplete information. Such an ability enables us to sort out logical from illogical or contradictory information. Humans experienced in field studies can synthesize diverse and disparate field observations, thereby expanding the opportunity for further discovery.

The value of human presence in conducting field work will depend on the inclusion in crews of experienced scientists with relevant scientific judgment and intuition. Their participation is, however, insufficient if they are not given the opportunity to perform as scientists. For example, the plans, procedures, and schedules of geological traverses must be sufficiently flexible to allow scientist-astronauts to modify sampling procedures, time on site, traverse routes, and so on, on the basis of their real-time assessment of in situ observations. To restrict this flexibility is to relegate the scientist-astronaut to the role of a human robot controlled from Earth.

The discussion of the advantages of human presence in planetary exploration is not theoretical: it has been demonstrated on the Apollo lunar missions.[1] Twelve astronauts, in six missions of increasing complexity, conducted tasks ranging from surface sample collection, with associated observations and photographic documentation of the geological context, to drilling and coring of the regolith, to emplacement of geophysical instruments. Photographic documentation of the sample sites proved invaluable in the interpretation of analyses of the returned samples. The astronauts, despite being encumbered by the spacesuits, proved adept at dealing with unforeseen problems such as repairing their roving vehicle and wrestling stuck drill bits and core tubes out of the ground. The geological training of the crews and the (relayed) interaction with the science teams in the Houston "back room" were sufficiently good to prove that excellent science can be accomplished in human exploration. Although the last Apollo mission included a scientist, many of the potential advantages of his presence were negated by the short duration of the mission and its rigid timeline.

As illustrative examples of human exploration activities, four diverse applications are examined that are particularly enhanced by the techniques of field investigation. In no particular order, these are the study of the lunar regolith as a probe of solar history, the search for martian fossil and extant life, determination of the meteorite bombardment history of the inner solar system, and the study of martian climate history. It can obviously be argued that, in theory, any of the discussed field activities could be accomplished robotically given sufficient advances in robotics and an adequate budget. That possibility is not examined here; CHEX's sole purpose is to look at the more useful activities that human explorers might conduct given their presence on the Moon or Mars for reasons other than science.

The committee hastens to note that it does not expect that a few missions or so will provide sufficient data to yield final, definitive answers to the scientific problems addressed by the examples of field activities mentioned below. Field experience on Earth relevant to determining climate history and to the origin of life and Apollo experience pertaining to solar emission history and to deciphering cratering flux demonstrate the complexity as well as the potential of the challenge.

Unraveling Solar Particle Emission History

Knowledge of long-term variations in the properties of the solar wind and solar energetic particles could provide important clues about the evolution of the Sun and the role of the solar wind in the formation and early development of the solar system.[2] Because solar wind particles impinge on and are implanted in the Moon's regolith, it may be possible to measure these variations by analysis of carefully selected lunar samples with a known geological context. This selection entails establishing the age of a given subunit. We must understand the early growth, formational dynamics, and continued evolution of the regolith through time. Thus, this activity is a field study problem in both geology and solar physics.

Study of the early growth and formation of the regolith is best accomplished by a two-pronged approach. First, excavations into the regolith should be studied to provide detailed geological information on its three-dimensional structure. At mare sites, it should be possible to excavate (in trenches or pits) and/or core down to the local lava flow bedrock (at depths of 5 to 8 meters). In such a manner, researchers could study regolith-bedrock contacts and learn about the earliest stages of regolith growth, an area that is poorly understood.

Second, study of the incipient growth of regolith on fresh bedrock surfaces on the Moon (for example, melt sheets of large fresh craters) would provide data for making inferences about stages of early growth exposed in regolith-bedrock contacts elsewhere on the Moon.[3,4]

Both of these studies require detailed field work, not only to collect samples intelligently, but also to make the observations and synthesize the visual clues needed to understand regolith growth dynamics. Outcrops of bedrock, such as those discovered on the wall of Hadley Rille by the Apollo 15 astronauts, are logical sites to begin such explorations.

To obtain "snapshots" of the solar particle output in ancient times, we need to find ancient regoliths on the Moon. Such fossil regoliths might be found sandwiched between lava flows of radiometrically determinable age. Locating such deposits and selecting unaltered or minimally altered samples for laboratory analysis (to measure the chemical and isotopic properties of these precisely controlled samples) are complex tasks requiring field study. Data from a variety of sites will constitute a set of solar wind "index fossils," that is, detailed measurements of the chemical and isotopic properties of the Sun at precisely defined intervals in the geological past. These can then be used to interpret and understand the solar record preserved in the regolith all over the Moon. Such knowledge will enable scientists to better interpret the solar record at regolith trenches and pits that may be excavated at other sites on the Moon, for example, during the construction of an underground habitat or the emplacement of instruments.

The Search for Life on Mars

The search for potential fossil and extant life on Mars, however low the probability for its existence is thought to be, continues to be a substantial goal of Mars exploration.[5,6] Detailed field studies will be required for this search, using robots initially but with increasing proximate human participation as the capability develops. Indeed, the robotic search for evidence of life on Mars began with the Viking landers in 1976.

The identification of sites to be analyzed for traces of life will require both extensive and intensive studies. These will include preliminary sampling by machines and, probably, robotic sample return to Earth. Even on Earth, however, the environments occupied by organisms are diverse and not necessarily obvious: there are organisms that thrive or survive within rock surfaces, in association with thermal vents and hot springs, at ice-water interfaces, and in liquid inclusions in salt deposits.[7] Proper site selection therefore may be motivated to a considerable extent by subtle idiosyncrasies: a crust within a sediment bed, a discoloration on ice or rock, or a boundary film between permafrost and regolith. Site selection will require subjective decisions based on astute observations of the specific locale, probably requiring a trained field observer.

Additionally, access to important sites may require the versatility of human workers. For instance, complex maneuvers will be required to reach sites in the polar ice caps or in the canyons of Valles Marineris, and coring or drilling may be required to reach ice-regolith interfaces or geothermal zones.

Evidence for past or present life on Mars will probably be sought in at least three ways: macroscopic and microscopic imaging, isotopic and chemical analysis, and culturing suspected life forms. Imaging procedures are capable of detecting macroscopic remains (such as stromatolites) and microscopic fossils. Because of the unique character of biomolecules, chemical methods are by far the most sensitive methods available to identify life, past or present. Isotopic analysis of carbon-bearing (e.g., organics, carbonate) or inorganic (e.g., sulfur) deposits can provide evidence for life because biochemical reactions create distinct isotopic fractionations.

Each of these analytical methods requires highly sophisticated sample processing and instrumentation. Imaging to search for microfossils will require sample preparation and electron microscopy. Chemical analyses will require chromatographic separations and mass spectrometry. Isotopic analyses will rely on chemical processing and high-resolution mass spectrometry. Initially, samples should be returned to Earth for analysis. However, the subsequent search for life will probably require iterative field study and in situ analysis because of the need for rapid feedback between analysis and further sampling.

The sophistication of the analytical methods and the variability of sample types that must be anticipated weigh against full automation of such analyses in the foreseeable future; human field workers/laboratory technicians will be required. On the other hand, this required analytical sophistication and complexity could argue for continued sample return. CHEX anticipates that trade-offs between in situ analysis and sample return will have to be made on the basis of further experience with martian materials and development of microanalytical techniques.

Impact History of the Terrestrial Planets

Through the study of impact history, the geological time scale for the formation of the surface units of the terrestrial planets can be reconstructed.[8] This process involves understanding the flux history of impacting bodies and then using such knowledge to convert relative ages determined by the density of impact craters into the estimates of absolute age required to address such topics as geological evolution and biological history.[9]

Determining the history of the cratering bombardment flux for the planets is, in practice, difficult. It involves obtaining samples appropriate for isotopic age dating from a variety of geological settings and locations; one must be able to unambiguously relate such samples to geological features of known relative age. For the latest stages of planetary evolution on both the Moon and Mars, there exists a variety of volcanic plains, from which "grab" samples are likely to yield lava crystallization ages appropriate to interpret as extrusion ages for the flows. Thus absolute ages for large tracts of planetary surfaces can be determined rather directly. Selecting a variety of grab samples is easily accomplished through robotic means and was, in fact, accomplished on the Moon in the 1970s by the Luna 16, 20, and 24 missions of the former Soviet Union.

In the earliest phases of planetary history, most geological units consist of crater deposits. In contrast to determining the absolute ages of lava plains, the dating of impact features is rather difficult. The only samples appropriate for dating large impact craters are relatively clast-free samples of impact melt, which typically constitute a few percent of the ejecta in cratering events. Although traces of a crater impact melt sheet can be recognized remotely and robotic missions can retrieve samples from such locales, it is not certain that such samples will be appropriate for radiometric dating. Even if such samples yield analytically good ages, their interpretation and relation to the age of the impact crater remain problematical. The careful collection of geologically controlled samples for dating impact craters is a difficult and complex problem and can be aided by human decisions and interactions.

Martian Climate History

Extensive channel systems on Mars suggest a warmer and wetter climate in the past. Layered deposits visible in the polar ice caps may have preserved a unique record of climate swings that occurred over the last few hundred million years.[10] Portions of this record may be recovered by drilling into the "sediments" and ice and analyzing the core samples. The two major causes of climate variations are thought to be martian orbital effects and temporal changes in solar irradiance. Because orbital effects have periods in the range of 10^5 to 10^6 years, their signal might be determined by studying a statistically significant number of cycles. After extraction of the signal due to orbital effects, the remaining variations might reveal the solar effects. Comparison with similar terrestrial data may verify a common external forcing function for global climatic changes in planetary atmospheres. The martian atmosphere is in many ways a simpler system than the terrestrial atmosphere because of the absence of a biosphere and massive oceans. Sorting out orbital from solar effects on climate may therefore be done more easily for martian samples than for terrestrial ones.

Human participation in these experiments would have two advantages: human judgment is needed to locate the best sites for drilling, and the number of samples would probably be so large that it would be best to conduct at least some of the chemical, isotopic, and mineralogical analyses in situ rather than after return to Earth.

EMPLACEMENT AND ATTENDANCE OF LARGE OR COMPLEX INSTRUMENTS

The use of the Moon as a platform for continuing studies of the planets, the Sun, other astronomical objects, and cosmic rays is an intriguing possibility. Although many instruments could be emplaced robotically, improved results could come from human interaction through more accurate positioning and troubleshooting capability. In addition, larger and more complex instruments conceivably could be constructed with human intervention. To some extent, having humans nearby could expedite maintenance and repair of broken equipment.

For probing the properties and environments of the Moon and Mars, instruments such as seismometers and meteorological stations will be necessary. While rudimentary facilities can be deployed globally by robotic probes, careful emplacement and attendance of advanced instruments at a few sites by humans may enable more sophisticated measurements with greater accuracy and precision. For example, placing a seismometer squarely on bedrock provides good coupling to the planet and improves the quality of the data dramatically over its emplacement on loose rubble. In fact, humans have significant experience emplacing seismometers, including on the Moon and, via robot surrogates, on the ocean floor and Mars. With the establishment of martian meteorology stations, significantly advanced instrumentation could be emplaced by humans, including tall towers or active sounders such as lidars, which could profile the atmosphere in considerable detail.

The surface of the Moon represents, potentially, an excellent platform for selected astronomical studies.[11,12] The lack of any appreciable atmosphere allows distortion-free images and complete spectral coverage. Sites shielded from direct sunlight can use passively cooled infrared detectors, obviating the need for expendable cryogens. Early missions to the Moon could carry small telescopes, which could be emplaced robotically. However, studies have indicated that fully assembled telescopes with apertures of the order of 1 to 2 m are the largest that could be deployed on the Moon in the initial phases of a lunar exploration program.[13] Larger telescopes would require assembly in place, most likely with on-site human assistance.

Several examples of the types of astronomical observations CHEX believes to be appropriate for a lunar observatory are noted below. However, the committee cautions that there has not yet been an independent, systematic analysis of how one should plan for astronomical or space physics observations in conjunction with a program of human exploration. Indeed, studies sponsored by proponents look at the Moon essentially in isolation from alternative ways (for example, in Earth orbit or ground-based) of conducting the desired observations.[14,15] The report of the Synthesis Group, for example, discusses the possibility of establishing a magnetospheric observatory on the Moon.[16] However, spacecraft in other orbits around Earth might be far superior platforms for studies that use remote sensing techniques to study the global properties of the magnetosphere. Others have suggested that astronauts on the Moon set up and maintain an observatory for monitoring variations in the composition of the solar wind. Although the lunar surface is a good place to study the solar wind's long-term, integrated composition, experience from the Apollo program shows that local magnetic fields complicate and invalidate the study of any short-term variations from the lunar surface. Although much was learned about the solar wind from analysis of samples collected in aluminum foils deployed by the

Apollo astronauts, future studies of the solar wind's composition using collection techniques would be better performed from a free-flying spacecraft that can face the Sun at all times.[17]

The Astronomy and Astrophysics Survey Committee addressed lunar-based astronomy in its chapter, "Astronomy and the Space Exploration Initiative."[18] It recognized the potential to conduct some first-rate astronomy from the Moon, at the same time pointing out potential disadvantages as well as unknowns about the lunar environment that must be ascertained before one can properly evaluate the possibilities. As is true with planetary science, any program of lunar-based astronomy must be constructed in the context of a vigorous and comprehensive astronomy program with Earth-based and free-flying components.

The European Space Agency's recent Phase-1 study of science on and from the Moon also found specific opportunities for astronomical observations, especially interferometry. However, it too urged a conservative approach and recommended a set of further studies.[19]

CHEX endorses the findings of the Astronomy and Astrophysics Survey Committee report on the next decade in astronomy,[20] which called for an evolutionary approach to lunar astronomy, one that complements the Earth-orbiting and ground-based astronomy program. It urged that such a step-by-step approach incorporate a comparative analysis of different opportunities, assessment of the lunar environment, initiation of advanced technology and instrument development (both, as has already been mentioned, considerably underfunded in current NASA programs), and progressive use of certain new techniques first on Earth, then in Earth orbit, and finally on the Moon. The Survey Committee advocated early initiation of a suitable small automated lunar astronomy mission as a reasonable way to start.[21]

Detection and Study of Other Solar Systems

A major objective that can be addressed from the Moon is the detection and characterization of planetary systems around other stars.[22] This goal was endorsed by the Astronomy and Astrophysics Survey Committee[23] and in a recent National Aeronautics and Space Administration report.[24] A particularly powerful tool for such a search is a large optical or infrared interferometer. One approach is to use an array of five 1.5-m passively cooled telescopes that could be individually soft-landed on the Moon and put into operation with limited human intervention for observations in the 0.2- to 5-micron range.[25]

The Moon is potentially superior to Earth orbit for such a device because its gravity and solid surface (free from seismic disturbances) can stabilize interferometer baselines without the complex metrology and continuous station-keeping needed with free-flying telescopes. Proposals to use humans to construct and align such a large interferometer recognize the difficulty in trying to do so robotically.

Study of High-Energy Cosmic Rays

The energy spectrum of galactic cosmic rays is known to have a change in slope, or a knee, between 10^{15} and 10^{16} electron volts (eV).[26] Possible explanations for the knee include a decrease in the effectiveness of acceleration of particles by shocks or an increase in the leakage of the more energetic particles out of the galaxy. To distinguish between these and other possibilities, researchers need to know the variation of elemental abundances of the cosmic-ray particles both above and below the knee.[27] There are, however, no direct composition measurements near the knee, and estimates of the composition range from pure hydrogen to pure iron. A lunar site would be highly suitable for an experiment designed to make such measurements, the so-called High Energy Abundance Project (HEAP).[28]

The Moon is ideal because it has no atmosphere and the heaviest part of HEAP, more than 150 metric tons of inert absorbing material, could consist of lunar soil. These measurements are not possible from Earth because of the atmosphere, nor are they practical in Earth orbit because of the cost of transporting that necessary amount of material into space. The 4-m cube of layered detectors and soil is perhaps most easily constructed by robotically assisted humans rather than robots, and humans would probably need to perform occasional maintenance.

Advanced Pinhole Occulter

The study of high-energy processes both in the Sun and in cosmic sources requires subarc second imaging in corresponding high-energy emissions such as hard x rays and gamma rays.[29] At such energies, imaging by conventional techniques (such as mirrors and lenses) is not possible. The emissions can, however, be imaged using "pinhole-camera methods" such as coded aperture masks and pairs of parallel-slit grids, which produce a Moiré fringe pattern in the detector plane.[30] The requirement for a sufficiently large field of view sets lower limits on the characteristic dimension of the apertures (be they pinholes or slits), and in turn the angular resolution requirement sets a lower limit to the separation of the grid pairs or masks.

Instrumentation of this type with modest collecting area and angular resolution down to a few arc seconds has been considered for use in Earth orbit around the turn of the century. Advanced, second-generation (subarc second) instruments of this genre would require accurate and stable positioning of apertures some hundreds of meters apart, an apparent impracticality for orbiting structures. Such a goal might, however, be met by a large lunar-based structure, one that would be extremely stable to both translational and torsional deformation. On-site engineers might be required to construct such a structure to the necessary tolerances and to conduct maintenance operations such as realignment of apertures.

LIFE SCIENCES

One of the more important physical features that influenced the evolution of life on Earth, and which places constraints on the development and functioning of all living organisms, is gravity. Once the factor of gravity is removed from the environment, living systems are altered, and the study of such alterations may lead to new insights into life processes.

The space life sciences are still in their infancy, and there have been few opportunities to carry out well-controlled experiments on living organisms in space. Thus it is not yet possible to predict how prolonged exposure to near-zero or fractional gravity will alter living systems. However, sufficient information is available to know that the absence of normal gravity profoundly alters living systems; thus exploration missions to the Moon and Mars will offer additional opportunity beyond Earth-orbiting space stations, to investigate the fundamental biological processes by which gravity affects living organisms.[31]

Missions to the Moon and Mars will also provide an opportunity for behavioral studies on crews under highly stressful conditions as well as over prolonged periods of time in close confinement. Such research would build on more than three decades of experience of human behavior and performance gathered from overwintering personnel at polar research stations. However, behavioral studies of the crews at a lunar outpost or on a Mars mission will provide new insights into human behavior because no polar base or even space station environment can duplicate all the conditions astronauts would experience on extended mission in deep space.[32] In the case of Mars, additional stress will result from the absence of any ready means of escape.

Both the gravitational biology and the behavioral studies are truly opportunistic; they are not now currently of high scientific priority in the life sciences community absent a program of human space exploration.

SCIENCE ENABLED BY TECHNOLOGY DEVELOPED FOR A MOON/MARS PROGRAM

The technology developments needed for successful exploration of the Moon and Mars are numerous and are spread throughout many disciplines. For example, a recent study identified 14 relevant areas of technology development.[33] Some of the general benefits to scientific investigations of two of these areas—spacesuits and telerobotics—are discussed above.

Some technology developments could enable robotic space-science missions unrelated to Moon/Mars exploration. For example, nuclear electric propulsion could enable several high-priority missions in heliospheric physics. Principal among these is the so-called interstellar probe.[34] This mission would penetrate a significant distance beyond the heliopause to provide the first comprehensive in situ studies of the plasma, energetic particles, cosmic rays, magnetic fields, gas, and dust in interstellar space. An advanced propulsion system is required to send a spacecraft 250 astronomical units from the Sun in significantly less than the 25 years or more required by conventional propulsion aided by gravity assists. Once such an advanced propulsion system is available, it could also be used for other high-energy missions, such as to propel instruments to large distances above the solar poles or into a short-period, circular solar polar orbit, and, perhaps, even a short-period eccentric orbit that skims through the solar corona at altitudes as low as three solar radii.[35]

SCIENTIFIC COMMUNITY PARTICIPATION

CHEX has given considered thought to how space science might benefit from the existence of a program of human exploration of the Moon and Mars, undertaken primarily for reasons other than science. History tells us that no matter when such a program is undertaken, a major activity will be scientific research. Indeed, CHEX concludes that there will be opportunities offering the potential for significantly enhancing our understanding of the Moon and Mars and for using them selectively as observation platforms. CHEX thus foresees a productive scientific role for human explorers as well as for continuing and enhanced robotic missions. The obvious conclusion is that scientists must participate in any eventual program of human exploration, although the question of how best to involve them must still be answered.

Scientists' past experiences with piloted spaceflight have been both good and bad. We can learn much from those (particularly the Apollo program) in terms of how NASA should approach science management and the involvement of scientists in a program of human exploration. That topic is under study and will be the subject of the third CHEX report. It is already clear to the committee, however, that scientists must be intimately involved in every stage of the endeavor and contribute to success by assuring that quality science is accomplished, that the science supported takes the best advantage of human presence, and that the resources available to the whole of space science are competitively allocated.

REFERENCES

1. William David Compton, *Where No Man Has Gone Before, A History of the Apollo Lunar Exploration Missions*, The NASA History Series, NASA SP-4214, NASA, Washington, D.C., 1989.
2. NASA, *A Planetary Science Strategy for the Moon*, JSC-25920, Lunar Exploration Science Working Group, Johnson Space Center, Houston, Texas, July 1992.
3. NASA, *Geosciences and a Lunar Base: A Comprehensive Plan for Lunar Exploration*, NASA Conference Publication 3070, NASA, Washington, D.C., 1990.
4. NASA, *A Planetary Science Strategy for the Moon*, JSC-25920, Lunar Exploration Science Working Group, Johnson Space Center, Houston, Texas, July 1992, page 8.
5. Space Studies Board, *The Search for Life's Origins: Progress and Future Directions in Planetary Biology and Chemical Evolution*, National Academy Press, Washington, D.C., 1990, page 8.
6. Space Studies Board, *1990 Update to Strategy for the Exploration of the Inner Planets*, National Academy Press, Washington, D.C., 1990, page 24.
7. Space Studies Board, *Biological Contamination of Mars: Issues and Recommendations*, National Academy Press, Washington, D.C., 1992, Chapter 4.
8. Space Studies Board, *Strategy for Exploration of the Inner Planets: 1977-1987*, National Academy of Sciences, Washington, D.C., 1978, page 71.
9. NASA, *A Planetary Science Strategy for the Moon*, JSC-25920, Lunar Exploration Science Working Group, Johnson Space Center, Houston, Texas, July 1992, page 6.
10. Space Studies Board, *Space Science in the Twenty-First Century: Imperatives for the Decades 1995 to 2015—Planetary and Lunar Exploration*, National Academy Press, Washington, D.C., 1988, page 101.
11. Y. Kondo (ed.), *Observatories in Earth Orbit and Beyond*, Proceedings of the 123rd Colloquium of the International Astronomical Union, Greenbelt, Maryland, April 24-27, 1990, Kluwer Academic Publishers, Dordrecht, The Netherlands, 1990.
12. Astronomy and Astrophysics Survey Committee, *The Decade of Discovery in Astronomy and Astrophysics*, National Academy Press, Washington, D.C., 1991, Chapter 6.
13. Synthesis Group, *America at the Threshold*, Report of the Synthesis Group on America's Space Exploration Initiative, U.S. Government Printing Office, Washington, D.C., 1991, page A-24.
14. NASA, *Future Astronomical Observatories on the Moon*, NASA Conference Publication 2489, NASA, Washington, D.C., 1988.
15. Michael J. Mumma and Harlan J. Smith (eds.), *Astrophysics from the Moon*, AIP Conference Proceedings 207, American Institute of Physics, New York, 1990.
16. Synthesis Group, *America at the Threshold*, Report of the Synthesis Group on America's Space Exploration Initiative, U.S. Government Printing Office, Washington, D.C., 1991, page A-26.
17. Space Studies Board, *A Strategy for the Explorer Program for Solar and Space Physics*, National Academy Press, Washington, D.C., 1984, pages 29-30.
18. Astronomy and Astrophysics Survey Committee, *The Decade of Discovery in Astronomy and Astrophysics*, National Academy Press, Washington, D.C., 1991, Chapter 6.
19. European Space Agency, *Mission to the Moon: Europe's Priorities for the Scientific Exploration and Utilization of the Moon*, Report of the Lunar Study Steering Group, ESA SP-1150, European Space Agency, Noordwijk, The Netherlands, June 1992.
20. Astronomy and Astrophysics Survey Committee, *The Decade of Discovery in Astronomy and Astrophysics*, National Academy Press, Washington, D.C., 1991.
21. Astronomy and Astrophysics Survey Committee, *The Decade of Discovery in Astronomy and Astrophysics*, National Academy Press, Washington, D.C., 1991, page 108.

22. Bernard F. Burke, "Astrophysics from the Moon," *Science,* **250**, December 7, 1990, page 1365.

23. Astronomy and Astrophysics Survey Committee, *The Decade of Discovery in Astronomy and Astrophysics*, National Academy Press, Washington, D.C., 1991, page 104.

24. NASA, *TOPS: Toward Other Planetary Systems*, A report by the Solar System Exploration Division, NASA, Washington, D.C., 1992.

25. Astronomy and Astrophysics Survey Committee, *The Decade of Discovery in Astronomy and Astrophysics,* National Academy Press, Washington, D.C., 1991, page 104.

26. Space Studies Board, *Space Science in the Twenty-First Century: Imperatives for the Decades 1995-2015—Astronomy and Astrophysics,* National Academy Press, Washington, D.C., 1988, page 31.

27. Space Studies Board, *Assessment of Programs in Solar and Space Physics 1991,* National Academy Press, Washington, D.C., 1991, page 14.

28. Michael L. Cherry, "Particle Astrophysics and Cosmic Ray Studies from a Lunar Base," *Astrophysics from the Moon,* Michael J. Mumma and Harlan J. Smith (eds.), AIP Conference Proceedings 207, American Institute of Physics, New York, 1990, page 593.

29. Laurence E. Peterson, "High Energy Astrophysics from the Moon," *Astrophysics from the Moon,* Michael J. Mumma and Harlan J. Smith (eds.) AIP Conference Proceedings 207, American Institute of Physics, New York, 1990, page 345.

30. Paul Gorenstein, "High-Energy Astronomy from a Lunar Base," *Future Astronomical Observatories on the Moon,* NASA Conference Publication 2489, NASA, Washington, D.C., 1988, page 45.

31. Space Studies Board, *Assessment of Programs in Space Biology and Medicine 1991,* National Academy Press, Washington, D.C., 1991.

32. Space Studies Board, *Assessment of Programs in Space Biology and Medicine 1991,* National Academy Press, Washington, D.C., 1991, Chapter 4.

33. Synthesis Group, *America at the Threshold,* Report of the Synthesis Group on America's Space Exploration Initiative, U.S. Government Printing Office, Washington, D.C., 1991, page 83.

34. Space Studies Board, *Space Science in the Twenty-First Century: Imperatives for the Decades 1995 to 2015—Solar and Space Physics,* National Academy Press, Washington, D.C., 1988.

35. Space Studies Board, *Assessment of Programs in Space Biology and Medicine 1991,* National Academy Press, Washington, D.C., 1991, Chapter 4.

Bibliography

Committee on Human Exploration of Space, *Human Exploration of Space: A Review of NASA's 90-Day Study and Alternatives,* National Academy Press, Washington, D.C., 1990.

Committee on Space Policy, *Toward a New Era in Space: Realigning Policies to New Realities,* National Academy Press, Washington, D.C., 1988.

Space Science Board, *HZE-Particle Effects in Manned Spaceflight,* National Academy of Sciences, Washington, D.C., 1973.

Space Science Board, *Life Beyond the Earth's Environment: The Biology of Living Organisms in Space,* National Academy of Sciences, Washington, D.C., 1979.

Space Science Board, *Origin and Evolution of Life—Implications for the Planets: A Scientific Strategy for the 1980's,* National Academy of Sciences, Washington, D.C., 1981.

Space Science Board, *Post-Viking Biological Investigations of Mars,* National Academy of Sciences, Washington, D.C., 1977.

Space Science Board, *Recommendations on Quarantine Policy for Mars, Jupiter, Saturn, Uranus, Neptune, and Titan,* National Academy of Sciences, Washington, D.C., 1978.

Space Science Board, *Space Science in the Twenty-First Century: Imperatives for the Decades 1995 to 2015—Life Sciences,* National Academy Press, Washington, D.C., 1988.

Space Science Board, *Strategy for Exploration of the Inner Planets: 1977-1987,* National Academy of Sciences, Washington, D.C., 1978.

Space Science Board, *A Strategy for Space Biology and Medical Science for the 1980s and 1990s,* National Academy Press, Washington, D.C., 1987.

Space Studies Board, *1990 Update to Strategy for the Exploration of the Inner Planets,* National Academy Press, Washington, D.C., 1990.

Space Studies Board, *Assessment of Programs in Space Biology and Medicine 1991,* National Academy Press, Washington, D.C., 1991.

Space Studies Board, *Biological Contamination of Mars: Current Assessment and Recommendations,* National Academy Press, Washington, D.C., 1992.

Space Studies Board, *International Cooperation for Mars Exploration and Sample Return*, National Academy Press, Washington, D.C., 1990.

Space Studies Board, *The Search for Life's Origins: Progress and Future Directions in Planetary Biology and Chemical Evolution*, National Academy Press, Washington, D.C., 1990.

Appendix

Participating Discipline Committees

COMMITTEE ON SPACE BIOLOGY AND MEDICINE
FRED W. TUREK, Northwestern University, *Chair*
ROBERT M. BERNE, University of Virginia, Charlottesville
PETER DEWS, Harvard Medical School
R.J. MICHAEL FRY, Oak Ridge National Laboratory
FRANCIS (DREW) GAFFNEY, Southwestern Medical Center, Dallas
EDWARD GOETZL, University of California Medical Center, San Francisco
ROBERT HELMREICH, University of Texas, Austin
JAMES LACKNER, Brandeis University
BARRY W. PETERSON, Northwestern University
CLINTON T. RUBIN, State University of New York, Stony Brook
ALAN L. SCHILLER, Mt. Sinai Medical Center
TOM SCOTT, University of North Carolina, Chapel Hill
WARREN SINCLAIR, National Council on Radiation Protection and Measurements
WILLIAM THOMPSON, North Carolina State University, Raleigh
FRED WILT, University of California, Berkeley

COMMITTEE ON SOLAR AND SPACE PHYSICS*

MARCIA NEUGEBAUER, Jet Propulsion Laboratory, *Co-Chair*
THOMAS CRAVENS, University of Kansas
JONATHAN F. ORMES, Goddard Space Flight Center
GEORGE K. PARKS, University of Washington
DOUGLAS M. RABIN, National Optical Astronomy Observatories
DAVID M. RUST, Johns Hopkins University
RAYMOND J. WALKER, University of California, Los Angeles
YUK L. YUNG, California Institute of Technology
RONALD D. ZWICKL, National Oceanic and Atmospheric Administration

COMMITTEE ON SOLAR-TERRESTRIAL RESEARCH*

DONALD J. WILLIAMS, Applied Physics Laboratory, *Co-Chair*
ALAN C. CUMMINGS, California Institute of Technology
GORDON EMSLIE, University of Alabama
DAVID C. FRITTS, University of Colorado
ROLANDO R. GARCIA, National Center for Atmospheric Research
MARGARET G. KIVELSON, University of California, Los Angeles
DAVID J. McCOMAS, Los Alamos National Laboratory
JONATHAN F. ORMES, Goddard Space Flight Center
EUGENE N. PARKER, University of Chicago
JAMES F. VICKREY, SRI International

---

*The National Research Council's (NRC) Committee on Solar-Terrestrial Research (CSTR) and Committee on Solar and Space Physics (CSSP) meet jointly as a federated committee and report directly to their parent NRC boards, the Board on Atmospheric Sciences and Climate for CSTR and the Space Studies Board for CSSP.

COMMITTEE ON PLANETARY AND LUNAR EXPLORATION

LARRY W. ESPOSITO, University of Colorado, *Chair*
RETA BEEBE, New Mexico State University, Las Cruces
ALAN P. BOSS, Carnegie Institution of Washington
ANITA L. COCHRAN, University of Texas, Austin
PETER J. GIERASCH, Cornell University
WILLIAM S. KURTH, University of Iowa, Iowa City
LUCY-ANN McFADDEN, University of California, San Diego
CHRISTOPHER P. McKAY, NASA Ames Research Center
DUANE O. MUHLEMAN, California Institute of Technology
NORMAN R. PACE, Indiana University
GRAHAM RYDER, Lunar and Planetary Institute
PAUL D. SPUDIS, Lunar and Planetary Institute
PETER H. STONE, Massachusetts Institute of Technology
GEORGE WETHERILL, Carnegie Institution of Washington
RICHARD W. ZUREK, Jet Propulsion Laboratory

Science Management
in the
Human Exploration of Space

Science Management in the Human Exploration of Space

Committee on Human Exploration

Space Studies Board

Commission on Physical Sciences, Mathematics, and Applications

National Research Council

NATIONAL ACADEMY PRESS
Washington, D.C.   1997

NOTICE: The project that is the subject of this report was approved by the Governing Board of the National Research Council, whose members are drawn from the councils of the National Academy of Sciences, the National Academy of Engineering, and the Institute of Medicine. The members of the committee responsible for the report were chosen for their special competences and with regard for appropriate balance.

This report has been reviewed by a group other than the authors according to procedures approved by a Report Review Committee consisting of members of the National Academy of Sciences, the National Academy of Engineering, and the Institute of Medicine.

The National Academy of Sciences is a private, nonprofit, self-perpetuating society of distinguished scholars engaged in scientific and engineering research, dedicated to the furtherance of science and technology and to their use for the general welfare. Upon the authority of the charter granted to it by the Congress in 1863, the Academy has a mandate that requires it to advise the federal government on scientific and technical matters. Dr. Bruce Alberts is president of the National Academy of Sciences.

The National Academy of Engineering was established in 1964, under the charter of the National Academy of Sciences, as a parallel organization of outstanding engineers. It is autonomous in its administration and in the selection of its members, sharing with the National Academy of Sciences the responsibility for advising the federal government. The National Academy of Engineering also sponsors engineering programs aimed at meeting national needs, encourages education and research, and recognizes the superior achievements of engineers. Dr. William A. Wulf is president of the National Academy of Engineering.

The Institute of Medicine was established in 1970 by the National Academy of Sciences to secure the services of eminent members of appropriate professions in the examination of policy matters pertaining to the health of the public. The Institute acts under the responsibility given to the National Academy of Sciences by its congressional charter to be an adviser to the federal government and, upon its own initiative, to identify issues of medical care, research, and education. Dr. Kenneth I. Shine is president of the Institute of Medicine.

The National Research Council was organized by the National Academy of Sciences in 1916 to associate the broad community of science and technology with the Academy's purposes of furthering knowledge and advising the federal government. Functioning in accordance with general policies determined by the Academy, the Council has become the principal operating agency of both the National Academy of Sciences and the National Academy of Engineering in providing services to the government, the public, and the scientific and engineering communities. The Council is administered jointly by both Academies and the Institute of Medicine. Dr. Bruce Alberts and Dr. William A. Wulf are chairman and vice chairman, respectively, of the National Research Council.

Support for this project was provided by Contract NASW 4627 and Contract NASW 96013 between the National Academy of Sciences and the National Aeronautics and Space Administration. Any opinions, findings, conclusions, or recommendations expressed in this publication are those of the author(s) and do not necessarily reflect the views of the organizations or agencies that provided support for this project.

Cover: Mars mosaic image courtesy of the U.S. Geological Survey, Flagstaff, Arizona. Lunar crescent image courtesy of Dennis di Cicco. Cover design by Penny E. Margolskee.

Copies of this report are available free of charge from

Space Studies Board, National Research Council, 2101 Constitution Avenue, N.W., Washington, D.C. 20418

Copyright 1997 by the National Academy of Sciences. All rights reserved.

Printed in the United States of America

COMMITTEE ON HUMAN EXPLORATION

NOEL W. HINNERS, Lockheed Martin Astronautics, *Chair*
WILLIAM J. MERRELL, JR., H. John Heinz III Center
ROBERT H. MOSER, University of New Mexico
JOHN E. NAUGLE, National Aeronautics and Space Administration (retired)
MARCIA S. SMITH, Congressional Research Service

PETER W. ROONEY and MARC S. ALLEN, Study Directors
BARBARA L. JONES, Administrative Associate

SPACE STUDIES BOARD

CLAUDE R. CANIZARES, Massachusetts Institute of Technology, *Chair*
MARK R. ABBOTT, Oregon State University
JOHN A. ARMSTRONG,* IBM Corporation (retired)
JAMES P. BAGIAN, Environmental Protection Agency
DANIEL N. BAKER, University of Colorado
LAWRENCE BOGORAD, Harvard University
DONALD E. BROWNLEE, University of Washington
JOHN J. DONEGAN, John Donegan Associates, Inc.
GERARD W. ELVERUM, JR., TRW
ANTHONY W. ENGLAND, University of Michigan
DANIEL J. FINK,* D.J. Fink Associates, Inc.
MARTIN E. GLICKSMAN, Rensselaer Polytechnic Institute
RONALD GREELEY, Arizona State University
BILL GREEN, former member, U.S. House of Representatives
NOEL W. HINNERS,* Lockheed Martin Astronautics
ANDREW H. KNOLL, Harvard University
JANET G. LUHMANN, University of California, Berkeley
JOHN H. McELROY,* University of Texas, Arlington
ROBERTA BALSTAD MILLER, CIESIN
BERRIEN MOORE III, University of New Hampshire
KENNETH H. NEALSON, University of Wisconsin
MARY JANE OSBORN, University of Connecticut Health Center
SIMON OSTRACH, Case Western Reserve University
MORTON B. PANISH, AT&T Bell Laboratories (retired)
CARLÉ M. PIETERS, Brown University
MARCIA J. RIEKE, University of Arizona
JOHN A. SIMPSON, Enrico Fermi Institute
ROBERT E. WILLIAMS, Space Telescope Science Institute

MARC S. ALLEN, Director

---

*Former member.

COMMISSION ON PHYSICAL SCIENCES, MATHEMATICS, AND APPLICATIONS

ROBERT J. HERMANN, United Technologies Corporation, *Co-chair*
W. CARL LINEBERGER, University of Colorado, *Co-chair*
PETER M. BANKS, Environmental Research Institute of Michigan
LAWRENCE D. BROWN, University of Pennsylvania
RONALD G. DOUGLAS, Texas A&M University
JOHN E. ESTES, University of California, Santa Barbara
L. LOUIS HEGEDUS, Elf Atochem North America, Inc.
JOHN E. HOPCROFT, Cornell University
RHONDA J. HUGHES, Bryn Mawr College
SHIRLEY A. JACKSON, U.S. Nuclear Regulatory Commission
KENNETH H. KELLER, University of Minnesota
KENNETH I. KELLERMANN, National Radio Astronomy Observatory
MARGARET G. KIVELSON, University of California, Los Angeles
DANIEL KLEPPNER, Massachusetts Institute of Technology
JOHN KREICK, Sanders, a Lockheed Martin Company
MARSHA I. LESTER, University of Pennsylvania
THOMAS A. PRINCE, California Institute of Technology
NICHOLAS P. SAMIOS, Brookhaven National Laboratory
L.E. SCRIVEN, University of Minnesota
SHMUEL WINOGRAD, IBM T.J. Watson Research Center
CHARLES A. ZRAKET, MITRE Corporation (retired)

NORMAN METZGER, Executive Director

Foreword

From the dawn of the space age, human spaceflight and space science have made uneasy bedfellows. A 1960 report commissioned by Science Advisor George Kistiakowsky for President Eisenhower concluded that ". . . among the major reasons for attending the manned exploration of space are emotional compulsions and national aspirations. . . . It seems, therefore, to us at the present time that man-in-space cannot be justified on purely scientific grounds, although more thought may show that there are situations for which this is not true. On the other hand, it may be argued that much of the motivation and drive for the scientific exploration of space is derived from the dream of man's getting into space himself."[1] In addition to questions of motivation and justification, accommodating the frequently conflicting needs of human life support and scientific investigation inevitably increases pressures on finite financial and tangible resources.

The successes of joint crewed and scientific missions, from Apollo to the Hubble repair to Shuttle/MIR, show the possible benefits of cohabitation. Of course, there have also been periods of friction and consequently unrealized potential. This report of the Space Studies Board's Committee on Human Exploration examines U.S. spaceflight history and draws lessons about "best practices" for managing scientific research in conjunction with a human spaceflight program. Since NASA's current focus is the development and subsequent operation of a crewed orbital laboratory, the International Space Station, some of these lessons should be immediately useful. The report is intended to be especially germane for a national decision to resume human exploration beyond low Earth orbit.

Claude R. Canizares, *Chair*
Space Studies Board

---

[1]"Report of the Ad Hoc Panel on Man-in-Space," December 16, 1960, in *Exploring the Unknown, Volume I: Organizing for Exploration*, John W. Logsdon, ed., NASA SP-4407, NASA, Washington, D.C., 1995, p. 411.

Preface

In 1988 the National Academy of Sciences and the National Academy of Engineering stated in the report *Toward a New Era in Space: Realigning Policies to New Realities* that ". . . the ultimate decision to undertake further voyages of human exploration and to begin the process of expanding human activities into the solar system must be based on non-technical factors. It is clear, however, that if and when a program of human exploration is initiated, the U.S. research community must play a central role by providing the scientific advice necessary to help make numerous political and technical decisions."

Since its establishment in 1958, the Space Studies Board, formerly the Space Science Board, has been the principal independent advisory body on civil space research in the United States. In this capacity, the Board established the Committee on Human Exploration (CHEX) in 1989 to examine science and science policy matters concerned with the return of astronauts to the Moon and eventual voyages to Mars. The Board asked CHEX to consider three major questions:

1. What scientific knowledge is prerequisite for prolonged human space missions?

2. What scientific opportunities might derive from prolonged human space missions?

3. What basic principles should guide the management of both the prerequisite scientific research and the scientific activities that may be carried out in conjunction with human exploration?

This report addresses the third of these topics. The first was the subject of

Scientific Prerequisites for the Human Exploration of Space, published in 1993, and the second was treated in *Scientific Opportunities in the Human Exploration of Space,* published in 1994.

In developing principles to guide management of the science covered in the first two reports, the committee observed that the productivity of the scientific component of human space exploration appears to be correlated with the organizational approach and structure used to manage the program. It is reasonable, then, to look back and try to formulate principles and recommendations that can strengthen the prospects for future success. It was not the committee's charge or intent to tell NASA precisely how to organize itself; indeed, there are several possible organizational arrangements that would be consistent with the conclusions of this study. Moreover, no organizational arrangement can guarantee success in the absence of clearly articulated and commonly agreed on goals. Throughout its study, the committee has made a deliberate effort to find ways to abolish the historic dichotomy between space science and human exploration and to seek ways to encourage a synergistic partnership.

When the committee initiated its work in 1989, it appeared that NASA might proceed with a new initiative in the human exploration of the solar system, specifically human missions to the Moon and Mars, and there was an interest on the part of the Space Studies Board to influence these new activities. Since that time, urgency to proceed to an implementation phase abated as budget pressures and a drastically changed world political situation weighed against any near-term commitment. On the other hand, the nation's commitment to human presence in low Earth orbit has become firmer with the pending orbital assembly of the International Space Station. Moreover, interest in a Mars human exploration program has been aroused by the recent announcement of possible evidence of relic biological activity in a meteorite of martian origin. The associate administrators for space science and human exploration recently directed NASA field centers to initiate planning for an integrated approach that could be brought forward "sometime in the second decade of the next century." The fact that human exploration beyond low Earth orbit is once again a subject of public dialogue and active planning makes this report especially timely.

Noel W. Hinners, *Chair*
Committee on Human Exploration

Contents

EXECUTIVE SUMMARY 1

1 INTRODUCTION 5
Approach, 6
Management of the Classical (Robotic) Space Science Program, 7
A New Environment, 7
Notes and References, 8

2 PRINCIPLES FOR SCIENCE MANAGEMENT 10
Interaction between Space Science and Human Spaceflight Communities, 10
Management Principles, 19
References, 21

3 MANAGEMENT RECOMMENDATIONS 22
Science Prerequisites for Human Exploration (Enabling Science), 22
Science Enabled by Human Exploration, 26
Institutional Issues, 28
Notes and References, 32

BIBLIOGRAPHY 34

Executive Summary

Since the late 1960s, the post-Apollo future of human space exploration has been a subject of ongoing debate, incremental decisions, variable political support, ceaseless studies, and little progress or commitment toward a well-defined long-term goal. In 1989, President Bush attempted to establish a direction by announcing a long-term goal for the U.S. space program of returning humans to the Moon and then voyaging to Mars early in the 21st century. His proposal did not win political support. Indeed, implementation of human exploration of the solar system for a time virtually disappeared from public discussion, largely as a result of greatly increased federal budget pressures and the end of the Cold War, which in combination have brought about a de facto reprioritization of national goals, including an examination of the entire rationale for the U.S. civil space program.

Recently, steps have been taken to initiate integrated planning for the exploration of Mars. In parallel, the goals of the International Space Station (ISS) program include the conduct of life science research and the acquisition of practical operational experience needed to resolve issues related to long-duration human spaceflight. Concurrently, robotic exploration of the Moon and Mars is being pursued by the United States and other countries.

The Space Studies Board (SSB) constituted the Committee on Human Exploration (CHEX) in 1989 to examine the general question of the space science component of a future human exploration program. The first CHEX report, *Scientific Prerequisites for the Human Exploration of Space*,[1] addressed the question of what scientific knowledge is required to enable prolonged human space missions. The second CHEX report, *Scientific Opportunities in the Human Ex-*

ploration of Space,[2] addressed the question of what scientific opportunities might derive from prolonged human space missions.

During the development of these first two reports, it became evident to the committee that the mode of interaction between space science and human exploration has varied over the years, as evidenced by a succession of different NASA organizational structures. The committee reviewed the history of this interaction with the objective of developing a "lessons-learned" set of principles and recommendations for the future. The principles and recommendations thus evolved for managing the science component of a Moon/Mars program, whenever and however it is pursued, transcend political and administrative changes.

While this report is not intended to dictate precise organizational models, application of these principles and recommendations should facilitate a productive integration of science into a program of human exploration.

PRINCIPLES FOR SCIENCE MANAGEMENT

Three broad principles emerged from the committee's survey of past programs:

INTEGRATED SCIENCE PROGRAM—The scientific study of specific planetary bodies, such as the Moon and Mars, should be treated as an integral part of an overall solar system science program and not separated out simply because there may be concurrent interest in human exploration of those bodies. Thus, there should be a single Headquarters office responsible for conducting the scientific aspects of solar system exploration.

CLEAR PROGRAM GOALS AND PRIORITIES—A program of human spaceflight will have political, engineering, and technological goals in addition to its scientific goals. To avoid confusion and misunderstandings, the objectives of each individual component project or mission that integrates space science and human spaceflight should be clearly specified and prioritized.

JOINT SPACEFLIGHT/SCIENCE PROGRAM OFFICE—The offices responsible for human spaceflight and space science should jointly establish and staff a program office to collaboratively implement the scientific component of human exploration. As a model, that office should have responsibilities, functions, and reporting relationships similar to those that supported science in the Apollo, Skylab, and Apollo-Soyuz Test Project (ASTP) missions.

MANAGEMENT RECOMMENDATIONS

In addition to these broad principles, the committee developed a number of specific recommendations on managing space research in the context of a human exploration program. Divided into three general categories, these recommendations are as follows.

Science Prerequisites for Human Exploration (Enabling Science)

1. The program office charged with human exploration should establish the scientific and programmatic requirements needed to resolve the critical research and optimal performance issues enabling a human exploration program, such as a human mission to Mars. To define these requirements, the program office may enlist the assistance of other NASA offices, federal agencies, and the outside research community.

2. The scientific investigations required to resolve critical enabling research and optimal performance issues for a human exploration program should be selected by NASA's Headquarters science offices, or other designated agencies, using selection procedures based on broad solicitation, open and equitable competition, peer review, and adequate post-selection debriefings.

3. NASA should maintain a dedicated biomedical sciences office headed by a life scientist. This office should be given management visibility and decision-making authority commensurate with its critical role in the program. The option of having this office report directly to the NASA Administrator should be given careful consideration.

Science Enabled by Human Exploration

4. Each space research discipline should maintain a science strategy to be used as the basis for planning, prioritizing, selecting, and managing science, including that enabled by a human exploration program.

5. NASA's Headquarters science offices should select the scientific experiments enabled by a human exploration program according to established practices: community-wide opportunity announcements, open and equitable competition, and peer review.

6. The offices responsible for human exploration and for space science should jointly create a formal organizational structure for managing the enabled science component of a human exploration program.

Institutional Issues

7. Officials responsible for review of activities or protocols relating to human health and safety and planetary protection on human and robotic missions should be independent of the implementing program offices.

8. The external research community should have a leading role in defining and carrying out the scientific experiments conducted within a human exploration program.

9. A human exploration program organization must incorporate scientific personnel to assist in program planning and operations, and to serve as an interface between internal project management and the external scientific community. Such "in-house" scientists should be of a professional caliber that will enable them to compete on an equal basis with their academic colleagues for research opportunities offered by human exploration missions.

10. Working through their partnership in a joint spaceflight/science program office, the science offices should control the overall science management process, including the budgeting and disbursement of research funds.

REFERENCES

1. Space Studies Board, National Research Council, *Scientific Prerequisites for the Human Exploration of Space*, National Academy Press, Washington, D.C., 1993.

2. Space Studies Board, National Research Council, *Scientific Opportunities in the Human Exploration of Space*, National Academy Press, Washington, D.C., 1994.

1

Introduction

The post-Apollo future of human space exploration beyond Earth orbit has been a subject of ongoing debate and study with little progress or commitment toward a clearly defined long-term goal since the late 1960s. In 1989, President Bush attempted to establish direction by announcing a long-term goal for the U.S. space program of returning humans to the Moon and then voyaging to Mars early in the 21st century. His proposal, termed the Space Exploration Initiative (SEI), was not followed by political action, nor has it been pursued by the current Administration. There is continued support for U.S. leadership in an International Space Station (ISS) program, however, whose utilization relates directly to a goal of long-duration human spaceflight. Indeed, the Committee on Human Exploration's first report, *Scientific Prerequisites for the Human Exploration of Space*,[1] dealt specifically with the requirements for a microgravity research facility in space.

Recently, NASA's associate administrators for space science and human exploration issued a joint directive to the Jet Propulsion Laboratory and Johnson Space Center to form a multicenter working group to fully integrate robotic and human Mars exploration planning.[2] The integrated activity is intended to result in a proposal that can be brought forward for human exploration missions that could begin "sometime in the second decade of the next century."

The committee based its second report, *Scientific Opportunities in the Human Exploration of Space*,[3] on the assumption that any program of human exploration of the solar system would have significant science content; in fact, most exploration studies[4-9] depict science goals as major motivations for such a program. The November 1996 directive cited above specifically identifies "science planning and science strategy" as a focus area for the integrated planning effort.

Although no science requirement has been identified that can be met only by a human presence, the committee believes that the scientific community should take the initiative in determining what space science goals might benefit from a human spaceflight program, given that such a program exists primarily for other reasons.

In contemplating involvement with human flight programs, many space scientists are conditioned by the fact that, despite notable successes and benefits, interactions between the scientific and human spaceflight communities have sometimes been marked by friction and dubious accommodation. Both the successes and failures constitute important lessons for any future human exploration program; while preparing its reports on the enabling (prerequisite) and enabled (opportunistic) science[10] for a human exploration program, the committee recognized the value of reviewing the history of space science programs carried out within the larger context of a human exploration program. Thus, the committee and the Space Studies Board set out to determine what attributes of past programs, particularly management attributes, might minimize the conflict and maximize the potential for a productive integration of science with human exploration.

APPROACH

The committee identified several broad principles that have contributed to mission success in the past. In doing so the committee made use of histories by John Naugle,[11] Homer Newell,[12] and William Compton,[13] as well as the recollections and judgments of committee and Space Studies Board members, many of whom played major roles in the evolution of these principles. These inputs were augmented by views solicited from representatives of the current and past space science and human exploration program offices at NASA.

To aid in identifying the effects of different management structures and approaches, the committee first reviewed the history of space science programs conducted in the context of human exploration, including the robotic program that preceded Apollo. It then analyzed those programs that involved interactions between space science and human spaceflight in terms of where mission requirements were defined and where authority for experiment selection and responsibility for funding were vested. The resulting groupings are loosely referred to as management models, although they also happen to correspond to distinct eras in the evolution of NASA's programs. The committee also considered the historical development of space biomedicine—a disciplinary area identified in its *Prerequisites* report as critical to future human exploration programs. The committee then extracted lessons learned and developed some general principles that could be applied to future programs.

MANAGEMENT OF THE CLASSICAL (ROBOTIC) SPACE SCIENCE PROGRAM

During most of NASA's existence, the Office of Space Science (OSS)[14] has formulated, funded, and executed NASA's space science program. Advised by the Space Studies Board[15] and assisted by the scientific community, OSS established long-range objectives, devised missions, selected scientists to conduct experiments, and planned the data analysis program.[16] OSS funded all robotic missions, including those conducted to gather data in support of Apollo. It budgeted for the scientific instruments, the spacecraft, and the conduct of flight operations. Prior to the advent of the Space Shuttle, OSS budgeted for and procured the expendable launch vehicles used to launch NASA's spacecraft. (In recent times, the budget for expendable launch vehicles has been restored to OSS.) OSS selected a NASA field center to manage each mission, and that center appointed a project manager and a project scientist to implement the mission.

Policies and procedures for robotic space science missions emerged during the early days of the space program from a vigorous process in which the merits of alternative procedures were debated. In many cases, procedures used to manage successful scientific projects were generalized and incorporated into formal NASA policy. The approach adopted proved fruitful, especially in planetary exploration, but also in physics and astronomy. The scientific data that came from Ranger (ultimately), Surveyor, and Apollo; from planetary programs such as Mariner, Viking, Voyager, and Magellan; from space physics missions such as the Explorers, Pioneers, and Orbiting Solar Observatories, and from astronomy programs such as the Apollo Telescope Mount (ATM) on Skylab, the Orbiting Astronomical Observatory, the International Ultraviolet Explorer, the Compton Gamma Ray Observatory, the Cosmic Background Explorer, and the Hubble Space Telescope demonstrate the effectiveness of NASA's evolved policies and practices.

A NEW ENVIRONMENT

Those who created the structure to manage science in the Apollo program had a relatively clean slate to work with, but this will not be so in the future. Officials directing a future human exploration program will have to work within, or modify, deeply ingrained policies, procedures, and cultures built up by NASA and the scientific community over 40 years. In addition, NASA has entered into a cooperative research relationship with the National Institutes of Health, for example, which could play a role in gathering the enabling biomedical data needed to support extended space missions by humans. Similarly, future human exploration missions are likely to involve significant international collaboration, as does the ISS program today. As a consequence, participants external to NASA may play an increased role in structuring or implementing the program.

A future human exploration program is not likely to be a sprint to a single, scheduled event, as was the Apollo landing on the Moon. A more probable approach is a phased one using, perhaps, the "go as you pay" strategy recommended in the report of the Augustine Committee.[17] Indeed, the November 1996 directive provides that the requested planning proposal be "credible in all respects: technically, scientifically, fiscally, with respect to risk, etc."[18]

No management arrangement can substitute for effective leadership. Such leadership will be required to identify and resolve cultural and other conflicts that will likely arise in such a large, complex, and expensive endeavor as returning humans to the Moon or traveling to Mars. The International Space Station program offers an opportunity to experiment and to begin forging a consensus on an optimal management approach. This could lead to a closer integration of the science and human exploration communities than has been achieved in the past, with a commensurate increase in both the likelihood of a human exploration program and the ultimate scientific return from it.

NOTES AND REFERENCES

1. Space Studies Board, National Research Council, *Scientific Prerequisites for the Human Exploration of Space*, National Academy Press, Washington, D.C., 1993.

2. Letter to the directors of the Jet Propulsion Laboratory and Johnson Space Center from Associate Administrators Wilbur Trafton, Arnauld Nicogossian, and Wesley Huntress, November 7, 1996; a press release announcing a cooperative activity to jointly fund and manage two robotic missions to Mars due for launch in 2001 was issued on March 25, 1997: "Space Science and Human Space Flight Enterprises Agree to Joint Robotic Mars Lander Mission," NASA Release 97-51.

3. Space Studies Board, National Research Council, *Scientific Opportunities in the Human Exploration of Space*, National Academy Press, Washington, D.C., 1994.

4. President's Science Advisory Committee, Joint Space Panels, *The Space Program in the Post-Apollo Period*, U.S. Government Printing Office, Washington, D.C., February 1967.

5. National Aeronautics and Space Administration (NASA), *Beyond the Earth's Boundaries: Human Exploration of the Solar System in the 21st Century*, NASA, Washington, D.C., 1988.

6. National Aeronautics and Space Administration (NASA), *Leadership and America's Future in Space*, NASA, Washington, D.C., 1987.

7. National Aeronautics and Space Administration (NASA), *Report of the 90-day Study on Human Exploration of the Moon and Mars*, NASA, Washington, D.C., 1989.

8. Advisory Committee on the Future of the U.S. Space Program, *Report of the Advisory Committee on the Future of the U.S. Space Program* (the "Augustine report"), U.S. Government Printing Office, Washington, D.C., 1990.

9. Synthesis Group, *America at the Threshold*, Report of the Synthesis Group on America's Space Exploration Initiative, U.S. Government Printing Office, Washington, D.C., 1991.

10. Space Studies Board, National Research Council, *Scientific Prerequisites for the Human Exploration of Space*, National Academy Press, Washington, D.C., 1993; Space Studies Board, National Research Council, *Scientific Opportunities in the Human Exploration of Space*, National Academy Press, Washington, D.C., 1994.

11. John E. Naugle, *First Among Equals: The Selection of NASA Space Science Experiments*, NASA SP-4215, NASA, Washington, D.C., 1991, pp. 79-196.

12. Homer E. Newell, *Beyond the Atmosphere: Early Years of Space Science*, NASA SP-4211, NASA, Washington D.C., 1980.

INTRODUCTION

13. William D. Compton, *Where No Man Has Gone Before: A History of Apollo Lunar Exploration Missions*, NASA History Series, NASA SP-4214, NASA, Washington, D.C., 1989.

14. In the past, the office has been either the Office of Space Science (OSS) or the Office of Space Science and Applications (OSSA), depending on whether some or all of space applications, microgravity science, or life science were combined with space science. The report uses the OSS acronym in a general sense.

15. Formed in 1988 from the union of the Space Science Board and elements of the former Space Applications Board.

16. Office of Technology Assessment, *NASA's Office of Space Science and Applications: Process, Priorities, and Goals*, U.S. Government Printing Office, Washington, D.C., January 1992.

17. Advisory Committee on the Future of the U.S. Space Program, *Report of the Advisory Committee on the Future of the U.S. Space Program*, U.S. Government Printing Office, Washington, D.C., 1990, pp. 6, 28, and 48.

18. Letter to the directors of the Jet Propulsion Laboratory and Johnson Space Center from Associate Administrators Wilbur Trafton, Arnauld Nicogossian, and Wesley Huntress, November 7, 1996.

2

Principles for Science Management

Based on a review of the historical interactions between human spaceflight programs and the scientific community, the committee saw its challenge as establishing a set of principles that, when employed, could facilitate the productive integration of space science into a human exploration program. These principles, along with the more specific recommendations developed in Chapter 3, might serve as a guide for decisions on what science to do in conjunction with a human exploration program, how and when to bring the scientific community into the program, and how to define the responsibilities and authorities of participating NASA offices.

There has been significant evolution in the interaction between the space science and human spaceflight communities during NASA's 40-year history. The two communities have pushed and pulled until a workable accommodation was established for each program.[1] This history can be divided into three principal eras: early lunar exploration before the Apollo landings; the Saturn launcher-based programs (Apollo, Skylab, and the Apollo-Soyuz Test Project); and the post-Saturn Space Shuttle era. Each of these eras featured a distinct but evolutionary distribution of authorities and responsibilities among the science and human spaceflight offices.

INTERACTION BETWEEN SPACE SCIENCE AND HUMAN SPACEFLIGHT COMMUNITIES

Early Lunar Exploration

The management structure used during the Ranger, Surveyor, and Lunar Orbiter programs evolved over time, culminating in the structure used to manage the very successful Lunar Orbiter program. Unlike Lunar Orbiter, the Rangers and Surveyors were not initially conceived as support missions to human flight but were reoriented to this goal as the Apollo program progressed.

After the initial Soviet space successes with the Sputnik program in late 1957, the United States attempted to gain leadership in space exploration with several hasty, ill-conceived attempts to beat the Soviets to the Moon. These robotic missions either failed totally or reached the Moon after the Soviet missions. In 1959, NASA abandoned these crash programs and formulated a systematic program to explore the Moon and the nearby planets. Two challenging lunar programs, Ranger and Surveyor, were initiated. The first NASA spacecraft to be stabilized in all three axes, the first two Rangers were designed to explore the space environment between Earth and the Moon. Surveyor originally consisted of an orbiter and a soft lander, each carrying a variety of scientific instruments.

NASA Headquarters assigned both the Ranger and Surveyor projects to the Jet Propulsion Laboratory (JPL). When measured against the existing technology and the knowledge and experience of those involved, Ranger was probably the most difficult and certainly one of the most frustrating projects ever undertaken by the Office of Space Science (OSS). Each mission carried a number of scientific instruments, each to be furnished by a scientist, most of whom worked at universities or other government laboratories. To demonstrate that the spacecraft worked, the JPL Ranger project manager wanted to launch the first Ranger as soon as possible and viewed anything that stretched the schedule as an impediment to be eliminated. The scientists, believing that their experiments were the objective of the Ranger project, found themselves in conflict with JPL and frequently with each other. The project manager also found that he could not get reliable information about the performance of the Atlas-Agena launch vehicle under development by the Department of Defense, and he did not know how many instruments he could accommodate.

When NASA assigned responsibility to JPL to conduct the lunar and planetary program, the senior management of JPL argued that they needed their own scientific advisory structure to help them plan the program. They expected NASA Headquarters to approve the JPL program, send money, and then wait for the results. Responsible for the overall program and under pressure from Congress to beat the Soviets, however, NASA Headquarters chose not to delegate responsibility for formulation of the programs to its centers: JPL could conduct studies and make recommendations, but the final decisions would be made at Headquarters. Further, NASA money would be accompanied by technical directives that JPL must follow. There were also disagreements about who would select investigators, JPL or NASA Headquarters.

The first Ranger failed in August 1961, and five more failed before Ranger 7 transmitted back more than 4,000 pictures of the lunar surface in July 1964. Rangers 8 and 9, the last two, also succeeded, and returned more than 17,000 high-quality images of the lunar surface. A major share of Ranger's problems can be traced to the struggles of a new agency whose allocation of roles and responsibilities was still being established.[2]

Surveyor suffered delays and cost overruns. Atlas-Centaur, Surveyor's launch vehicle, failed on its first launch. In mid-1962, NASA eliminated the Surveyor orbiter and all of the lander's scientific instruments except those needed to fulfill Apollo requirements. The cancellation of the Surveyor orbiter created pressure on OSS to develop an alternative lunar orbiter, because the Office of Manned Space Flight (OMSF) needed lunar photographs to select Apollo landing sites. The Space Science Board urged Congress to fund a lunar orbiter, which it did. In October 1962, in response to OMSF requirements and congressional pressure, OSS and OMSF formed a joint working group to plan a Lunar Orbiter program to map the lunar surface. Since JPL was already saturated with Ranger and Surveyor, as well as the Mariner project, this working group was asked to select a NASA center to develop the Lunar Orbiter. In early 1963, OSS started the Lunar Orbiter program at the Langley Research Center.

Surveyor 1 landed on the Moon on June 2, 1966. Two months later, on August 10, 1966, Lunar Orbiter 1 returned its first pictures of the lunar surface. Five of the seven Surveyors succeeded, and all five Lunar Orbiters successfully completed their missions.

By the time of their successful missions, the primary purpose of all three of these programs was to provide information that the Apollo project needed. In the Lunar Orbiter project, OMSF, which had overall responsibility for the Apollo program, had a customer-like relationship with OSS. That is, OMSF expressed its requirements to OSS and left it to OSS to obtain the needed data within specified time constraints. Although OSS formulated and oversaw the development and operation of all three programs and took responsibility for delays and overruns, the customer model is not an exact representation because OSS sought and maintained funding for these missions as well. After early problems, the management approach evolved to successfully support the Apollo program and enable ground-breaking lunar science.

Several observations concerning the management of space science emerged from the experience of the early days of NASA's lunar exploration program. The chances of mission success are enhanced if the objectives of each specific project or mission are clearly specified. If the prime objective of a project is to gather engineering data on a new space system, for example, and the accomplishment of scientific experiments is a secondary objective, then that fact should be made clear to the scientists participating in the project. If the prime objective of the mission is to accomplish a scientific task, then that fact must be made equally clear to the project team, which should be judged by its success in accomplishing the scientific objectives of the mission, as well as by meeting schedule and budget commitments. Also, scientific goals can be pursued most effectively if conducted within the framework of a single space science program run by one NASA

Headquarters office, and leaving selection of investigators to the Headquarters science office.

Apollo, Skylab, Apollo-Soyuz

The interactions between science and human spaceflight in Apollo, Skylab, and the Apollo-Soyuz Test Project (ASTP) were much more complicated than those in the Lunar Orbiter case. Apollo began strictly as a human spaceflight mission, as NASA's initial plans included no scientific experiments. But the Space Science Board's 1962 Iowa Summer Study examined the role for the human in research on the lunar surface,[3] and the Physics Committee, an OSS advisory group, proposed that the astronauts place optical corner reflectors on the Moon. Ultimately, a substantial lunar research program arose from these suggestions and from other experiments proposed by other NASA advisory groups.

In the early 1960s, tension arose about the conduct of lunar science on Apollo—should there continue to be one lunar science program formulated by OSS or should the Apollo project formulate and conduct its own lunar science program? How should the science program be defined and funded, and by whom?

In March 1962, an ad hoc working group on Apollo lunar science was set up at the request of OMSF.[4] The ad hoc working group met three times in early 1962 and submitted a report to the Iowa Summer Study held that summer. In the fall of 1962, the associate administrator of OSS moved to set up a more formal Joint Working Group on scientific lunar exploration and the development of scientific experiments for Apollo, structured to report to both OSS and OMSF. Discussions between OSS and OMSF continued in 1963, leading in July 1963 to a reorganization of the Joint Working Group into the Manned Space Science Division, which continued to report to the two offices (Figure 2.1). Selection and preliminary development of experiments were assigned to OSS, and development of flight hardware and integration to OMSF; each office bore the costs for its share of the experiment development.

In September 1963, OMSF established a Manned Spaceflight Experiments Board to review all experiments, whether scientific experiments proposed by OSS, technology experiments proposed by NASA's Office of Advanced Research and Technology, or military experiments proposed by the Department of Defense (Figure 2.1 shows the NASA spaceflight organization at this time, including the Manned Space Science Division). The Manned Spaceflight Experiments Board examined the technical requirements of the experiments, such as the weight, orientation, and amount of power and astronaut time required. This board did not question the scientific merits of the scientific experiments that had been approved by the associate administrator for OSS, but, in his capacity as chairman of the board, the associate administrator for OMSF retained final approval authority for all experiments that flew on the Apollo missions. Some scientists believed that

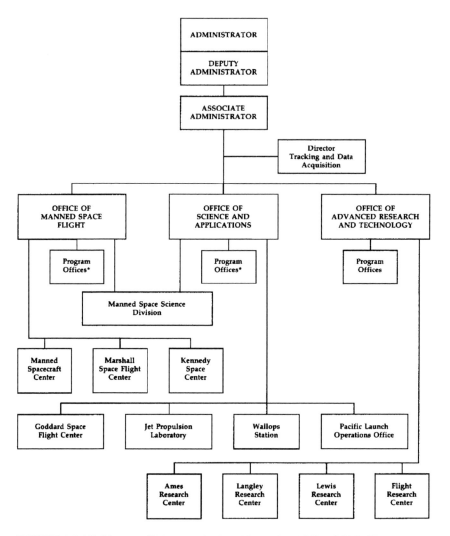

FIGURE 2.1 NASA spaceflight organization, November 1963. SOURCE: Reprinted from W.D. Compton, *Where No Man Has Gone Before: A History of Apollo Lunar Exploration Missions,* NASA History Series, NASA SP-4214, NASA, Washington, D.C., 1989.

this encroached on the OSS role in science selection and constituted an unneeded administrative burden.

In September 1965, the deputy administrator of NASA issued a directive allocating responsibility for aspects of manned spaceflight programs. In part, it confirmed the existing OSS-OMSF agreement and provided that OMSF would have the responsibility of developing scientific experiments selected by OSS. It also provided that OMSF would fund the experiments. While disagreements between the staffs of OSS and OMSF did not disappear, this established the principle that there would be one space science program formulated by OSS.

This imperfect arrangement continued until September 1967, when a NASA reorganization promoted OSS head Homer Newell to NASA Associate Administrator and made Newell's former deputy, Edgar Cortright, deputy associate administrator of OMSF. Under Newell's oversight, OSS and OMSF shortly thereafter created a joint Apollo Lunar Exploration Office to be staffed jointly by OMSF and OSS and to be physically and organizationally located in the OMSF Apollo Program Office (Figure 2.2). A former OSS manager of the highly successful Lunar Orbiter program was designated the new director of lunar exploration. Reporting to him were four assistant directors, all experienced OSS program managers and program scientists. For administrative matters, hardware development, and funding status, the director of lunar exploration reported to the director of the Apollo program office, but for all scientific matters he reported to the associate administrator for OSS. Thus, the Lunar Exploration Office was established not as a liaison office or working group, but rather as an integral component of the Apollo program organization within OMSF, charged with responsibility for lunar experiment hardware that would both meet the Apollo schedule and satisfy OSS science requirements.

The arrangement proved successful, based on several important factors. Cortright, now deputy associate administrator of OMSF, knew and trusted the OSS people in the Lunar Exploration Office and hence could assure the director of the Apollo program office that they would accomplish the tasks assigned to them. At the same time, because the leaders of the office were all experienced OSS employees, they enjoyed the confidence and support of the associate administrator of OSS and knew that when they had completed their work for the Apollo program they would return to OSS. Key to success were the shared recognition by OSS and OMSF of the need for a joint office and staffing of this office with experienced individuals of acknowledged achievement.

Having proved its worth during the Apollo missions, the joint project office concept was also applied to the Apollo Telescope Mount (ATM), the solar observatory operated by the astronauts on the Skylab space station (1973-1974). OSS had already selected experiments for the Advanced Orbiting Solar Observatory (AOSO) prior to approval of Skylab. Forced to cancel AOSO because of a shortage of funds, OSS transferred the instruments to the ATM. The ATM project manager reported jointly to the associate administrator for OSS and the associate administrator for OMSF, just as in the case of Apollo.

For the 1975 Apollo-Soyuz Test Project (ASTP), the associate administrator for OSS was given the responsibility for selecting the experiments to be performed during the mission. After a false start resulting from a desire to expedite

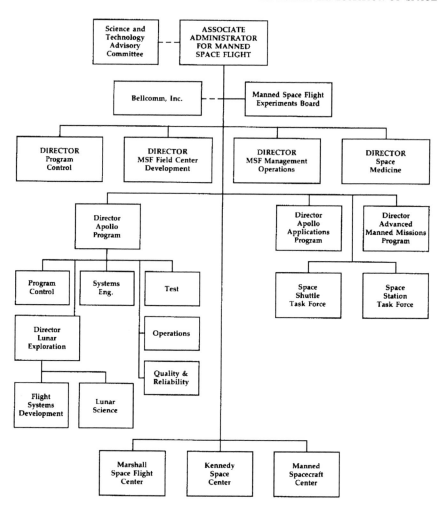

FIGURE 2.2 Office of Manned Space Flight organization, 1969. SOURCE: Reprinted from W.D. Compton, *Where No Man Has Gone Before: A History of Apollo Lunar Exploration Missions*, NASA History Series, NASA SP-4214, NASA, Washington, D.C., 1989.

selections, OSS assembled science working groups that successfully carried out a standard, if greatly accelerated, competitive selection process in just two months.

Several lessons were learned from Apollo, Skylab, and ASTP about the conduct of scientific research during human spaceflight. The formation of a joint program office, staffed by representatives of both NASA's science and human spaceflight offices, was shown to be an effective solution to the day-to-day tensions that arose between advocates of human exploration and advocates of science concerning the scientific experiments conducted during the Apollo missions. In addition, unifying Apollo and ASTP science objectives and processes with ongoing science management processes of OSS avoided duplication of activities, provided more effective cross-fertilization among scientific disciplines, and minimized confusion among policymakers. In one sense, however, the Apollo management approach reversed the earlier Lunar Orbiter approach: in Apollo, funds were sought and obtained for the science program by OMSF rather than by OSS, even though OSS selected the investigations to be carried out.

Shuttle/Spacelab

The relationship between the science and human spaceflight offices shifted again in the Space Shuttle/Spacelab program.

After the completion of the Apollo Moon landings, lack of support for an expensive space program in the Administration and Congress closed the Saturn-Apollo production lines and led NASA to propose a new, low-cost launch vehicle, the reusable Shuttle, for transporting humans to and from Earth orbit. In 1969-1970, NASA hoped to develop the Shuttle and a space station in parallel. Financial guidelines imposed on NASA by the Administration, however, precluded simultaneous development of two major human spaceflight systems. In 1971, when the members of NASA's Space Station Task Force found that the station had been postponed indefinitely, they abandoned work on it and, instead, turned to a pressurized, habitable container that the Shuttle could carry to and from orbit. Spacelab resulted from the work of the task force as a substitute for a continuously orbiting space station. In January 1972, the President approved the Shuttle program. In December 1972, the European Space Research Organization undertook to develop and manufacture Spacelab.[5]

There were disagreements within NASA and within the scientific community itself over the value of Shuttle/Spacelab. Scientists from disciplines that required long-duration observations or collection of data from orbits beyond those achievable by the Shuttle argued that a switch by NASA to the Shuttle/Spacelab system would leave them unable to conduct their research. Astronomers who had been disappointed by the loss of the first Orbiting Astronomical Observatory, on the other hand, were concerned about a national commitment to the Large Space Telescope (ultimately the Hubble Space Telescope) without a provision for the ability to repair any malfunctions.

Within NASA, the associate administrator of OSS organized and co-chaired the Shuttle Payload Planning Steering Group. This group, made up of members of OSS and OMSF, worked to make sure that OMSF, which was developing the Shuttle, understood space science requirements and that OSS understood the capabilities and constraints of the Shuttle. Out of these discussions emerged agreement on the need for upper stages for the Shuttle to place some scientific missions in higher orbits and a commitment by NASA that the Shuttle would be designed so that it would be able to launch and service the Hubble telescope, and that the Hubble telescope, in turn, would be able to be launched and serviced by the Shuttle.

After establishing the scientific requirements, the associate administrator for OSS controlled all space science and life science payload activity during development and operation of the Spacelab. The associate administrator for OSS funded and managed their development, rather than have them funded by OMSF and managed by a joint OSS/OMSF Spacelab Program Office as was done in Apollo and Skylab. The NASA administrator directed the associate administrator for OSS to select the scientists (the "payload specialists") who would fly on the Shuttle and conduct experiments in the Spacelab. The associate administrator for OSS would also direct the activities of the Spacelab Payload Project at the Marshall Space Flight Center and select the final payload complement.

This arrangement was nearly the opposite of that used for Lunar Orbiter. In the latter, the office responsible for human spaceflight set requirements for the science office for the data it needed to land humans on the Moon. In the case of the Shuttle/Spacelab program, the science office established "requirements" for the human spaceflight office to optimize the platform for science utilization. In reality, the fundamental characteristics of the Shuttle system were fixed by a complex network of budgetary, technological, and national security constraints, rather than being defined by scientific users. The resulting Shuttle capabilities were presented to the scientific community as an "opportunity" that could be adapted to a certain extent and exploited, for example by the Spacelab (and later Spacehab and other systems).

During development, testing, and operation of Spacelab, OSS continued to control the payload activity. When Spacelab became operational, OSS continued to fund and manage the development of all space and life science payloads. OSS selected not only the scientific investigators, but also the scientists who flew as payload specialists in the Shuttle to conduct experiments.

In spite of very high costs, greater than expected complexity, and initial skepticism of the science community, Shuttle/Spacelab has been successful in that some high-quality laboratory science has been accomplished. In addition, the Shuttle has been successfully used to repair and service the Solar Maximum Mission and the Hubble Space Telescope, as well as subsequently to upgrade the scientific capabilities of the Hubble. Several lessons were learned from experience with this program. First, science carried out within the context of human spaceflight needs the involvement of scientists at all stages of the program's conceptualization, development, and operation. This continuing involvement is necessary to ensure that realistic science goals are established that take advantage of human presence, and that missions, flight hardware, and procedures are designed to promote the accomplishment of science. In addition, the Spacelab program again confirmed the practice of the investigators being chosen by the science office rather than the program office responsible for flying the mission. In the Shuttle/Spacelab era, OSS budgeted for and managed science funding, similar to the earlier Lunar Orbiter but in contrast to Apollo. On the other hand, the Shuttle/Spacelab program reversed the customer relationship of Lunar Orbiter in the sense that OSS expressed accommodation requirements to OMSF, rather than OMSF tasking OSS with its data needs.

MANAGEMENT PRINCIPLES

In summary, a structure that grew out of the debate during the formulation of the Ranger and Surveyor programs was successfully used for the Lunar Orbiter program of robotic spacecraft that provided data used to select landing sites for the Apollo crews. During the early lunar exploration era, the office responsible for human spaceflight set requirements for the space science office in the sense that they told the science office what information they needed and when they needed it. The science office was given the management and budgetary authority to obtain needed data as it saw fit, albeit within a strict schedule. A more elaborate structure evolved during the Apollo program, and subsequently the Skylab and ASTP programs, to explicitly manage the interaction between the space science and human spaceflight programs. During this era, a joint management team that included representatives of the science office and the spaceflight office oversaw the conduct of space science within the context of the larger exploration programs. A third structure evolved during the era of the Spacelab program of pressurized modules and unpressurized pallets flown in the cargo bay of the Shuttle. During this period, the team approach that proved so successful during Apollo was largely abandoned, and the earlier model whereby the spaceflight office set requirements for the science office was essentially reversed, with the science office developing and negotiating requirements for orbital platforms to be designed, built, and launched by the spaceflight office.

The direction of the "customer-provider" relationship, and the related issue of which party advocates and obtains the science funding, are important because of their impact on project implementation. This in turn bears on the importance of clear priorities and the organizational locus of science decision making.

The committee identified three broad principles in its survey of the history of the interaction between space science and human spaceflight. Experience with the Ranger and Apollo programs demonstrates that waste and duplicated effort are minimized, and clear lines of authority are delineated, if the scientific aspects of solar system exploration are the responsibility of a single Headquarters office. Thus, the first principle is the following:

INTEGRATED SCIENCE PROGRAM—The scientific study of specific planetary bodies, such as the Moon and Mars, should be treated as an integral part of an overall solar system science program and not separated out simply because there may be concurrent interest in human exploration of those bodies. Thus, there should be a single Headquarters office responsible for conducting the scientific aspects of solar system exploration.

A common problem in programs with multiple goals is the relationship between actual and perceived priority of those goals. This issue has arisen to varying degrees in almost all of NASA's human-related space projects. Human exploration is not undertaken primarily for scientific reasons, but it has important scientific elements.[6] Thus it is essential that the relative priority of all the competing goals be well understood by all participants. Accordingly, the second broad principle is that clear objectives and priorities should be established at the level of individual component flight projects in the program in order to properly integrate science goals with the nonscience goals of human exploration:

CLEAR PROGRAM GOALS AND PRIORITIES—A program of human spaceflight will have political, engineering, and technological goals in addition to its scientific goals. To avoid confusion and misunderstandings, the objectives of each individual component project or mission that integrates space science and human spaceflight should be clearly specified and prioritized.

Although a human exploration program cannot be justified by scientific considerations alone, such missions have the potential, as noted in the committee's second report,[7] to provide significant scientific opportunities. NASA's experience indicates that the scientific return can be enhanced if there are good communications and a cooperative working relationship between engineering implementers and the scientists. A demonstrated means of facilitating productive integration of space science and human spaceflight is to establish a joint office. Thus, a third broad principle is that space science conducted in the context of a human exploration program should be managed through a joint spaceflight and science program office:

JOINT SPACEFLIGHT/SCIENCE PROGRAM OFFICE—The offices responsible for human spaceflight and space science should jointly establish and staff a program office to collaboratively implement the scientific component of human exploration. As a model, that office should have responsibilities, functions, and reporting relationships similar to those that supported science in the Apollo, Skylab, and Apollo-Soyuz Test Project (ASTP) missions.

Chapter 3 considers these principles and their implications in further detail in the context of the committee's two earlier reports.

REFERENCES

1. For more details, see Homer E. Newell, *Beyond the Atmosphere: Early Years of Space Science*, NASA SP-4211, NASA, Washington, D.C., 1980.
2. R. Cargill Hall, *Lunar Impact: A History of Project Ranger*, NASA SP-4210, NASA, Washington, D.C., 1977.
3. Space Science Board, *A Review of Space Research: The Report of the Summer Study Conducted Under the Auspices of the Space Science Board of the National Academy of Sciences at the State University of Iowa*, Iowa City, Iowa, June 17-Aug. 10, 1962, Publication 1079, National Academy of Sciences, Washington, D.C., 1962.
4. Details of the evolution of the relationship between science and the Apollo program are provided in *Where No Man Has Gone Before: A History of Apollo Lunar Exploration Missions*, by William D. Compton (NASA History Series, NASA SP-4214, NASA, Washington, D.C., 1989); see also *Beyond the Atmosphere: Early Years of Space Science*, by Homer E. Newell (NASA History Series, NASA SP-4211, 1980).
5. Douglas R. Lord, *SPACELAB, an International Success Story*, NASA SP-487, NASA, Washington, D.C., 1987.
6. Space Studies Board, National Research Council, *Scientific Prerequisites for the Human Exploration of Space*, National Academy Press, Washington, D.C., 1993.
7. Space Studies Board, National Research Council, *Scientific Opportunities in the Human Exploration of Space*, National Academy Press, Washington, D.C., 1994.

3

Management Recommendations

In its first (*Prerequisites*) report,[1] the committee designated the research required to undertake and optimize human exploration as "enabling science." In addition to enabling science, there is scientific research that can be conducted or significantly facilitated by the existence of a human exploration program. In its second (*Opportunities*) report,[2] the committee called this "enabled science" because it is enabled by the existence of the human exploration program. There is also a third category of space science, the classical space science conducted by the Office of Space Science (OSS) that does not involve humans working in space. This third category of science is straightforwardly managed according to well-established OSS policies and procedures similar to standard practices of the non-NASA research community, without the national policy issues and complicating effects of human presence.[3]

SCIENCE PREREQUISITES FOR HUMAN EXPLORATION (ENABLING SCIENCE)

The central issue in enabling science for a human exploration program concerns the collection and analysis of the prerequisite life science and biomedical data required in order to determine whether long-duration human spaceflight, such as that required for a voyage to Mars, is advisable or even possible. The committee's *Prerequisites* report identified two broad categories of enabling science required for undertaking human exploration of the inner solar system.

"Critical research issues" are those where present-day ignorance is great enough to pose unacceptably high risks to human spaceflight beyond low Earth orbit. These issues have the highest probability of being life-threatening or seriously debilitating to space explorers[4] —that is, they are in effect potential "showstoppers" for a human exploration mission.

A second category, "optimal performance issues," includes those that do not appear to be seriously detrimental to the health and well-being of humans in space, but that could degrade the performance of humans in flight or on extraterrestrial surfaces. Some of the issues in this category could later be found to be critical, especially in the areas of long-duration extraterrestrial habitation or return to terrestrial gravity following extended flight. In addition, some optimal performance issues relate to the enhancement of scientific yield.

Continued pursuit of enabling science research is required to determine whether human exploration of the solar system is, in fact, feasible. Much research related to the necessary objectives is already under way in various parts of NASA's organization.

Establishing Requirements for Enabling Science

The program office responsible for carrying out a human exploration program should be responsible for establishing the mission-critical enabling requirements. Program life scientists should be tasked with generating specific, goal-oriented questions to address anticipated problems in, for example, human physiology, psychology, and radiation protection. The program office cannot, however, be expected to possess the expertise necessary to fully develop all of the requirements alone, and experts without previous experience with NASA life science programs might contribute untapped expertise to the critical problems posed by long-duration human spaceflight. Program officials should call on other elements of NASA, for example, the office(s) responsible for the various space sciences, as well as non-NASA entities, such as the National Institutes of Health and the Department of Energy, for specialized assistance. Exploration program research goals should also be brought to the attention of recognized experts in the relevant disciplines within the academic community.

Thus, the committee recommends that:

1. The program office charged with human exploration should establish the scientific and programmatic requirements needed to resolve the critical research and optimal performance issues enabling a human exploration program, such as a human mission to Mars. To define these requirements, the program office may enlist the assistance of other NASA offices, federal agencies, and the outside research community.

Selection of Enabling Science Investigations

Once goal-oriented questions have been defined, the talents of the very best scientists and engineers will be necessary to obtain and analyze the data needed to satisfy these requirements. The U.S. civil space science program has achieved its many successes in creating new knowledge by developing, early in the space era, and continuing to refine a comprehensive, broadly based, widely understood and accepted investigator selection process based on peer review. Fundamental characteristics of this process have been described in several Space Studies Board reports.[5,6] The committee recommends that:

2. *The scientific investigations required to resolve critical enabling research and optimal performance issues for a human exploration program should be selected by NASA's Headquarters science offices, or other designated agencies, using selection procedures based on broad solicitation, open and equitable competition, peer review, and adequate post-selection debriefings.*[7]

The best medical scientists should participate in and review the enabling biomedical research programs.

Management of Space Biomedical Sciences

In carrying out Recommendation 2, it must be recognized that several factors complicate biomedical sciences in NASA. At times in the past, NASA management and the astronaut corps have perceived biomedical scientists as overly cautious. Early in the space program, for example, physicians responsible for the safety of humans in space argued for more data and more animal flights. This position conflicted with that of the managers of human spaceflight activities and the astronauts, who were anxious to orbit a man before the Soviets and were willing to accept greater risks.

More generally, NASA has had trouble engaging the interest of the highest-caliber biomedical scientists to conduct space-related research because the frontiers of biomedicine have been in terrestrial laboratories, rather than in space. Although over its three decades space biomedicine has had some significant spinoffs that have contributed to terrestrial medicine, such as the telemetering of data and miniaturization of equipment, the unique microgravity environment of space has not attracted the attention of the majority of researchers studying the physiology or diseases of Earth-bound humans. Even in discipline areas with particular promise for space-based research, the administrative and engineering complexity and long time scales of space experimentation tend to discourage investigators immersed in the broader world of fast-paced biological research. The main rationale for space-related biomedical research, then, has been the postulate that humans will spend extended periods in the space environment and explore the solar system.

In this context, many biomedical scientists have maintained that biomedical science should reside in a separate office with its own associate administrator (a life scientist).[8] The space biomedical sciences programs were maintained under the direction of the Office of Space Science (OSS) until late 1970 when NASA Headquarters decided that the only life science research, other than exobiology, that should be continued was research related to the safety of astronauts during lengthy spaceflights. OSS phased out its bioscience program, and the Office of Manned Space Flight (OMSF) was assigned responsibility for the remaining life science program. Skylab became the only facility for life science research, and the associate administrator of the OMSF selected the life science experiments conducted there. This arrangement prevailed from 1971 through 1975. In 1975, with Skylab completed, no human flights scheduled until the late 1970s, and no long-duration human flights scheduled for the foreseeable future beyond that, control of the total life science program was transferred back to OSS, where it remained until 1993.

When biomedical research was a component of OMSF there was a perceived inherent conflict of interest between purely scientific dictates and the imperative to get on with spaceflight. When incorporated into OSS, on the other hand, biomedical sciences tended to have lower priority relative to the traditional space physical sciences. Nonetheless, for most of NASA's history, its administrator, after examining the arguments, has rejected the notion of a separate office. Thus, until recently, space biomedicine has always been a subcomponent of either OSS or OMSF.

In 1993, the life sciences other than exobiology and studies related to the origin of life were transferred to the new Office of Life and Microgravity Sciences and Applications (OLMSA). Although it did not lead to a totally separate biomedical life sciences office, the policy of uniting the space biomedicine and microgravity sciences in one office recognized their broad similarities as experimental rather than observational sciences, and their similar infrastructure requirements as laboratory-oriented research disciplines. Advantages of this unification, which include strengthened management focus, have been compared with disadvantages in the Space Studies Board report *Managing the Space Sciences*.[9]

As a result of a sweeping policy-level review, which evaluated NASA's management structure in the context of a customer service model, NASA Administrator Daniel Goldin subsequently aggregated the agency's functional offices into "strategic enterprises." Initially, OLMSA, which has responsibility for space biomedicine, was grouped with the physical space sciences in the Scientific Research Enterprise. Later, OLMSA was relocated out of this enterprise, and joined with the Office of Space Flight (OSF) in the Human Exploration and Development of Space (HEDS) strategic enterprise.[10] Superficially, this configuration resembles the former management arrangement whereby the life sciences were included within a NASA program office whose main interest and responsibility were the conduct of spaceflight. But within HEDS, OLMSA's charter is defined as leadership in "space biological, physical, and chemical research and aerospace medicine, supporting technology development, and applications using the attributes of the space environment."[11] In addition, OLMSA's Research and Analysis (R&A) and flight programs are managed by customary peer-review practices to achieve broad scientific goals laid out in widely circulated solicitations.

In 1996, however, budgetary control over the scientific components of the space station program, including the NASA-Mir Research Program and Space Station Facilities and Utilization, was removed from OLMSA and placed under the management of the International Space Station program within OSF, OLMSA's partner in the HEDS strategic enterprise. In this arrangement, these important elements of the space laboratory research program are effectively once again vested in NASA's human spaceflight office, at least from a budgetary point of view, where they are directly subordinated to the priorities of the flight program.[12]

As argued in the Space Studies Board reports cited above (including the 1970 report), a program of extended-duration human spaceflight will place major new demands on the life sciences. In order to overcome past management problems, to bring additional high-quality research and researchers into the program, to ensure that those scientists are able to conduct cutting-edge research, and to enable NASA management to incorporate human biomedical factors directly into programmatic decisions at the highest levels, the committee recommends that:

3. NASA should maintain a dedicated biomedical sciences office headed by a life scientist. This office should be given management visibility and decision-making authority commensurate with its critical role in the program. The option of having this office report directly to the NASA Administrator should be given careful consideration.

SCIENCE ENABLED BY HUMAN EXPLORATION

Early examinations of enabled science in human exploration included the Space Science Board's Iowa Summer Study[13] on the scientific opportunities arising from the Apollo program, and the work of NASA's Task Force on the Scientific Uses of a Space Station.[14] In its *Opportunities* report, the present committee discussed the distinction between enabling and enabled science in human exploration. If these research categories are clearly distinguished and the distinction maintained during the course of implementation, then the most problematic issue that remains is the relative role of humans and robots. The tension between advocates of human exploration and advocates of robotic science missions has existed for a long time. Some researchers are convinced that space science objectives can be met using Earth-controlled or autonomous robotic spacecraft alone. Others believe, equally firmly, that the future viability of the entire U.S. civil space program depends on human presence in space. If these differences are carried into the future, the committee believes that the only result will be the diminution of the total U.S. space effort, probably at a significant cost to both groups.[15]

Humankind is still in the earliest phases of the exploration of the inner solar system. Further evolution can be expected in the concepts and details of a continuing program and in possibilities for enabled scientific research. Enabled science should be competitively evaluated in terms of its relationship to other space science initiatives and opportunities. Such an evaluation would involve not only scientific quality but also programmatic issues such as cost, schedule, and the value added by human presence. Cost is a particularly difficult issue to address. It is often argued that the incremental cost of individual science investigations is low in comparison to the total cost in human flight programs, and that such investigations therefore should be incorporated into the human flight mission. In the past, this rationale, combined with a flight opportunity, has been used to justify the flight of experiments whose merit was questionable or at least not clearly established by peer review. A pernicious side effect of this reasoning can be the imposition on the program or flight system of research requirements, together with their attendant costs and risks, that are unwarranted by the quality of the potential science return.

At the same time, there will arise occasions where it is decided, after thorough evaluation, that an investigation of high scientific merit should be accomplished within the human exploration program even though some programmatic considerations, such as cost, might argue for implementation through a purely robotic program. A past example illustrates this point: the Apollo Telescope Mount on Skylab successfully accomplished scientific objectives derived from planning for the robotic Advanced Orbiting Solar Observatory, a program that had been canceled in the space science program for budgetary reasons.

Space Science Strategies and Science Goals and Priorities

A key element in the conduct of space science has been the development of a research strategy for each major scientific discipline.[16] These strategies are developed, to the extent possible, without regard to the mode of implementation and evolve as knowledge, technology, and instrumentation advance. The strategies are crafted in such a way as to leave technical implementation to the agency programmatic planning process since the scientific committees that develop them are not constituted to have the information and expertise necessary to address implementation options in detail. Another reason that the research strategies avoid implementation recommendations is that they are intended to remain valid for 5 to 10 years, while the programmatic context changes on a much shorter time scale due to dynamics of annual budgets and overall national policy.

Each discipline's science strategy is used by NASA to help establish priorities for missions supporting that discipline. Because these priorities should apply also to research enabled by human exploration of the inner solar system, the committee recommends that:

4. *Each space research discipline should maintain a science strategy to be used as the basis for planning, prioritizing, selecting, and managing science, including that enabled by a human exploration program.*

Selection of Enabled Science Investigations

The overall merit of enabled space science is of central importance. Thus the decision-making process leading to the selection of a given enabled science project can only be articulated and defended by rigorous and systematic evaluation. Such tools already exist in the form of the practices and procedures used to select NASA's science programs.[17] In addition, there are good reasons for locating control of the selection process at NASA Headquarters.[18] As in the case of enabling science (Recommendation 2, above), rather than develop new procedures, the committee recommends that:

5. *NASA's Headquarters science offices should select the scientific experiments enabled by a human exploration program according to established practices: community-wide opportunity announcements, open and equitable competition, and peer review.*

Implementation of Enabled Science

Once science investigations are selected for a human exploration program, their actual implementation in the context of a specific set of mission constraints, e.g., mass, volume, and power requirements, necessarily involves interactions between the science offices and those charged with implementing the flight program. Taking note of the broad success of the procedures devised for this purpose during the Apollo, Skylab, and Apollo-Soyuz programs (see Chapter 2), the committee recommends that:

6. *The offices responsible for human exploration and for space science should jointly create a formal organizational structure for managing the enabled science component of a human exploration program.*

INSTITUTIONAL ISSUES

Protocol Review

In its first report, the committee commented that the potential hazards to be faced by the crews on human exploration missions beyond Earth orbit "must be adequately addressed within the context of a comprehensive program of health and safety. To do otherwise imposes unacceptable risks on the entire human exploration enterprise."[19] Experience from previous NASA programs, however, shows that concerns about astronaut health and safety, or about the forward and backward contamination of planetary bodies (planetary protection), can conflict with, or impede accomplishment of, the objectives of a specific mission. Analysis of the history of planetary quarantine during the Apollo era, for example, exposes a series of organizational and implementation problems, ranging from unclear allocation of authority and responsibility to deficient integration of engineering requirements and personnel training into the program.[20] One study concluded, after examining alternatives, with a preference that "a life science program office would be established within NASA with responsibilities for life science research and for protecting against extraterrestrial contamination, both outbound and inbound. As recommended in the 1960 NASA report, this program would carry status equivalent to that of other program offices within NASA."[21] Experience illustrates a clear need for independent objective review of the handling of these concerns and of constituent protocols by individuals and offices *not* responsible for the conduct of the flight program. The committee recommends that:

7. Officials responsible for review of activities or protocols relating to human health and safety and planetary protection on human and robotic missions should be independent of the implementing program offices.

The Role of Universities

Since its earliest days, the space program has benefited from the involvement of academic scientists in the development of science priorities, mission concepts, and instruments for spacecraft, and analysis of results. NASA needs the ideas, skills, and support of the academic community. This participation provides a steady source of new talent and rapid dissemination of results of the space program into the scientific and engineering communities.[22] In the early days of NASA, as competition for room on satellites increased, NASA established a formal procedure to ensure equitable access to its missions by all scientists whether at universities, NASA field centers, or other federal and commercial laboratories.[23] The human exploration of space will extend over a long period and thus will require a continual input of new talent. In addition, the program will generate new knowledge and technology. Therefore, the committee recommends that:

8. The external research community should have a leading role in defining and carrying out the scientific experiments conducted within a human exploration program.

This recommendation is consistent with an earlier Board recommendation that NASA's research be conducted out-of-house wherever possible.[24]

The Role of Scientific Expertise Within the Program

In the early days of NASA, many academic scientists and NASA engineers thought that scientific research should be conducted by academic scientists and that the function of NASA field centers should be to provide launch vehicles, spacecraft, and engineering help to these academics. It rapidly became apparent to both groups, however, that each NASA field center responsible for a NASA scientific mission, including the contractor-operated Jet Propulsion Laboratory, needed a group of highly qualified space scientists to help on a day-to-day basis with conceiving new missions, developing approved missions, and providing a channel of communication between the center and the academic community. The best way to guarantee and monitor the competence of these in-house scientists is to expect them to compete successfully with their academic colleagues for the opportunity to participate in the NASA space science program as investigators themselves.

In response to downsizing pressures and an agency desire to preserve and enhance the vitality of its science programs, the role of government space scientists, especially those at NASA field centers, has recently been reexamined in a number of Board studies and reports.[25-27] An alternate approach to the vital functions performed by these scientists that is structured around external, but tightly coupled, "science institutes" has been examined recently by NASA.[28] While not directed at a human exploration program, these analyses' rationale and conclusions apply directly to such a program, adapted perhaps to NASA's organizational configuration at such a time. The key point is that the functions currently exercised by NASA in-house project scientists are essential ones that should be maintained in any alternative organizational arrangement that might be adopted.

Consistent with findings of these studies, the committee makes the general recommendation that:

9. A human exploration program organization must incorporate scientific personnel to assist in program planning and operations, and to serve as an interface between internal project management and the external scientific community. Such "in-house" scientists should be of a professional caliber that will enable them to compete on an equal basis with their academic colleagues for research opportunities offered by human exploration missions.

Funding for Science in a Program of Human Exploration

The question of the programmatic source of funding for scientific experiments was considered at length by the committee. It can be argued that to attain the desired control over the science part of the program, the science office should budget for the science and control the science budget allocation and accountability process. A counter-argument holds that the expense of conducting science in conjunction with human spaceflight is so high that it should be budgeted under the human spaceflight program, where it would be a comparatively small cost element, to prevent it from crowding out other science priorities in science office budgets.

Historically, as Chapter 2 recounts, both of these approaches have been used at different times. During the early lunar exploration program, the Office of Space Science budgeted for the robotic missions. In contrast, the Apollo, Skylab, and Apollo-Soyuz programs themselves budgeted for their associated science. During Shuttle/Spacelab, the science programs have budgeted for science experiments and data analysis, although the Shuttle program has funded most of the integration of the payloads into the Shuttle and much of the common support equipment.

The committee concluded that there has been no clear correlation between science effectiveness and the programmatic source of science funding; rather, the committee's deliberations revealed that science effectiveness is correlated with the control of the management processes by which the science is selected and implemented. Science budgeting responsibility, on the other hand, has historically been largely a function of expediency and opportunity. For example, the high national priority of the Apollo program supported the high cost of Apollo science. In the Shuttle era, the assignment of science budgeting to the science office was driven by the desire to minimize the apparent cost of the Shuttle program.

It was pointed out to the committee that the synchronization of budgeting by the science office (or offices) in support of science enabled by a human exploration program remains a problem. That is, if the science office assumed the responsibility for budgeting human exploration program science, it would be required to ask for funds to plan and support science for human flight programs not yet approved in order for the science to be incorporated into the program in its early phases. This additional science funding could prove difficult to attract under these circumstances, and the science office would naturally be cautious about committing any of its existing resources specifically to such support. At the same time, NASA would like to be able to offer any scientific advantages of a human exploration program as part of its advocacy for that program. The committee appreciates the problem and suggests that the best approach is implied by Recommendation 4, that is, that strategic science planning that avoids prescribing implementation details can constitute a sound basis for preparation, negotiation, and participation, and offers the best assurance of appropriate balance and optimum synergy between robotic and human exploration. This approach would use the science strategies to inform a continuing dialogue and integration with the human exploration enterprise, strengthening both efforts and helping forestall late and ineffective science involvement.

In a zero-growth or declining budget environment, such as exists now and which also existed as Apollo tailed off, one cannot pretend that the higher cost of doing business within human spaceflight programs has no impact on science programs (see, for example, note 12 below). Thus, while scientific accomplishment does not appear to have been strongly correlated with the source of funding in the past, control of the science budgets by the science offices may, in fact, be essential to maintain the quality of the research program and a productive balance with flight system development in the future. The committee's general principle favoring the establishment of a joint spaceflight/science program office provides a mechanism for this within the context of a sound management structure; the committee therefore recommends that:

10. Working through their partnership in a joint spaceflight/science program office, the science offices should control the overall science management process, including the budgeting and disbursement of research funds.

NOTES AND REFERENCES

1. Space Studies Board, National Research Council, *Scientific Prerequisites for the Human Exploration of Space*, National Academy Press, Washington, D.C., 1993, pp. 10-12.

2. Space Studies Board, National Research Council, *Scientific Opportunities in the Human Exploration of Space*, National Academy Press, Washington, D.C., 1994, pp. 6-7.

3. Office of Technology Assessment, *NASA's Office of Space Science and Applications: Process, Priorities, and Goals*, U.S. Government Printing Office, Washington, D.C., January 1992.

4. Space Studies Board, National Research Council, *Radiation Hazards to Crews of Interplanetary Missions: Biological Issues and Research Strategies*, National Academy Press, Washington, D.C., 1996.

5. Space Studies Board, letter report to NASA Associate Administrator Harry Holloway from Louis J. Lanzerotti and Fred W. Turek, April 26, 1993.

6. Space Studies Board, National Research Council, *Managing the Space Sciences*, National Academy Press, Washington, D.C., 1995; Recommendations 5-1 through 5-8, pp. 57-58.

7. These considerations have most recently been addressed by the Board in connection with the Explorer mission line. See the report *Assessment of Recent Changes in the Explorer Program*, Space Studies Board, National Research Council, National Academy Press, Washington, D.C., 1996.

8. These views can be traced back to early days of the human flight program. See, for example, the Space Science Board's *Life Sciences in Space*, National Academy of Sciences, Washington, D.C., 1970, p. 19.

9. Space Studies Board, National Research Council, *Managing the Space Sciences*, National Academy Press, Washington, D.C., 1995, p. 37.

10. National Aeronautics and Space Administration, *NASA Strategic Plan, February 1995*, NASA, Washington, D.C.

11. National Aeronautics and Space Administration, *Budget Estimates—Fiscal Year 1998*, p. SAT 2-3, 1997. Scientific objectives are also prominent in the four top-level HEDS goals presented in *NASA's Enterprise for the Human Exploration and Development of Space—The Strategic Plan, January 1996*, NASA, Washington, D.C.

12. This is clearly indicated by supporting narrative in NASA's FY98 *Budget Estimates* volume: "This past year NASA consolidated the management of Space Station research and technology, science utilization, and payload development with the Space Station development and operations program in order to enhance the integrated management of the total content of the annual $2.1 billion budget. The Space Station program manager is now responsible for the cost, schedule, and technical

MANAGEMENT RECOMMENDATIONS

performance of the total program. The OLMSA and OMTPE [Office of Mission to Planet Earth] remain responsible for establishing the research requirements to be accommodated on the space station and will respond to the direction of the program manager to ensure the utilization priorities and requirements are consistent with the overall Space Station objectives." (p. HSF 1-2)

13. Space Science Board, National Research Council, *A Review of Space Research: The Report of the Summer Study Conducted Under the Auspices of the Space Science Board of the National Academy of Sciences at the State University of Iowa, Iowa City*, Iowa, June 17-Aug. 10, 1962, Publication 1079, National Academy of Sciences, Washington, D.C., 1962.

14. Task Force on the Scientific Uses of Space Station, *Space Station Summer Study Report—March 1985*, NASA, Washington, D.C., March 21, 1985.

15. The relationship between human and robotic exploration is discussed at greater length on pp. 9-15 in the Space Studies Board report, *Scientific Opportunities in the Human Exploration of Space*.

16. See, for example, Space Studies Board, National Research Council, *An Integrated Strategy for the Planetary Sciences: 1995-2010*, National Academy Press, Washington, D.C., 1994, and Space Studies Board, National Research Council, *Strategy for Space Biology and Medical Science for the 1980s and 1990s*, National Academy Press, Washington, D.C., 1987.

17. The Space Studies Board addressed aspects of NASA research selection procedures in several letter reports: letter to Associate Administrator Harry Holloway on April 26, 1993 (*Space Studies Board Annual Report*—1993, p. 32); letter to Life Science Division Director Joan Vernikos on July 26, 1995 (*Space Studies Board Annual Report*—1995, p. 85).

18. Space Studies Board, National Research Council, *Managing the Space Sciences*, National Academy Press, Washington, D.C., 1995, pp. 57-58.

19. Space Studies Board, National Research Council, *Scientific Prerequisites for the Human Exploration of Space*, National Academy Press, Washington, D.C., 1993, p. 12.

20. These problems are briefly referred to in the Space Studies Board report *Mars Sample Return—Issues and Recommendations*, National Academy Press, Washington, D.C., 1997, p. 35. A detailed account of the Apollo quarantine program experience is provided in *Back Contamination: Lessons Learned During the Apollo Lunar Quarantine Program*, by John R. Bagby, Jr., July 1, 1975 (prepared for the Jet Propulsion Laboratory under Contract #560226).

21. T. Mahoney, *Organizational Strategies for the Protection Against Back Contamination*, NASA-CR-149274, Final Report, University of Minnesota, St. Paul, Minn., 1976, pp. 39 and 47. This recommendation provides a sample organization chart that shows an Office of Life Science reporting, in parallel with the Office of Space Science and the Office of Manned Space Flight (and several others), directly to an Associate Administrator for Programs. The "1960 NASA report" cited in the quotation was the report of the NASA Bioscience Advisory Committee, dated January 25, 1960. In his report (cited in note 19 above), J. Bagby concluded that "[m]anagement of any future quarantine operation should be established as a special program office out of the office of the Administrator of NASA" (p. 42, emphasis in the original).

22. Homer E. Newell, *Beyond the Atmosphere: Early Years of Space Science*, NASA History Series, NASA SP-4211, NASA, Washington, D.C., 1980, pp. 223-241.

23. John E. Naugle, *First Among Equals: The Selection of NASA Space Science Experiments*, NASA History Series, NASA SP-4215, NASA, Washington, D.C., 1991, pp. 79-196.

24. Space Studies Board, National Research Council, *Managing the Space Sciences*, National Academy Press, Washington, D.C., 1995, pp. 43-44.

25. Space Studies Board, letter to NASA Chief Scientist France A. Cordova from Claude R. Canizares, March 29, 1995, *Space Studies Board Annual Report—1995*, p. 74.

26. Space Studies Board, letter to NASA Chief Scientist France A. Cordova from Claude A. Canizares, August 11, 1995, *Space Studies Board Annual Report—1995*, p. 88.

27. Space Studies Board, National Research Council, *Managing the Space Sciences*, National Academy Press, Washington, D.C., 1995.

28. National Aeronautics and Space Administration, *NASA Science Institutes Plan*, NASA, Washington, D.C., 1996.

Bibliography

Advisory Committee on the Future of the U.S. Space Program, *Report of the Advisory Committee on the Future of the U.S. Space Program*, U.S. Government Printing Office, Washington, D.C., 1990.

Committee on Human Exploration of Space, National Research Council, *Human Exploration of Space: A Review of NASA's 90-Day Study and Alternatives*, National Academy Press, Washington, D.C., 1990.

Committee on Space Policy, National Academy of Sciences-National Academy of Engineering, *Toward a New Era in Space: Realigning Policies to New Realities*, National Academy Press, Washington, D.C., 1988.

NASA, *Report of the 90-Day Study on Human Exploration of the Moon and Mars*, NASA, Washington, D.C., 1989.

NASA Advisory Council, *Exploring the Living Universe: A Strategy for Space Life Sciences*, Report of the NASA Life Sciences Strategic Planning Study Committee, NASA, Washington, D.C., 1988.

NASA Advisory Council, *Strategic Considerations for Support of Humans in Space and in Moon/Mars Exploration Missions*, Vol. I & II, Life Sciences Research and Technology Programs, Aerospace Medicine Advisory Committee, NASA, Washington, D.C., 1992.

National Commission on Space, *Pioneering the Space Frontier*, The Report of the National Commission on Space, Bantam Books, New York, 1986.

National Council on Radiation Protection and Measurements, *Guidance on Radiation Received in Space Activities*, NCRP Report No. 98, National Council on Radiation Protection and Measurements, Bethesda, Maryland, 1989.

Office of Exploration, *Beyond the Earth's Boundaries: Human Exploration of the Solar System in the 21st Century*, NASA, Washington, D.C., 1988.

Office of Exploration, *Leadership and America's Future in Space*, NASA, Washington, D.C., 1987.

Office of Space Science and Applications, *Cardiopulmonary Discipline Science Plan*, Life Sciences Division, NASA, Washington, D.C., 1991.

Office of Space Science and Applications, *Controlled Ecologic Life Support Systems (CELSS)*, Life Sciences Division, NASA, Washington, D.C., 1991.

Office of Space Science and Applications, *Developmental Biology Discipline Plan*, Life Sciences Division, NASA, Washington, D.C., 1991.
Office of Space Science and Applications, *Human Factors Discipline Science Plan*, Life Sciences Division, NASA, Washington, D.C., 1991.
Office of Space Science and Applications, *Musculoskeletal Discipline Science Plan*, Life Sciences Division, NASA, Washington, D.C., 1991.
Office of Space Science and Applications, *Neuroscience Discipline Science Plan*, Life Sciences Division, NASA, Washington, D.C., 1991.
Office of Space Science and Applications, *Regulatory Physiology Discipline Plan*, Life Sciences Division, NASA, Washington, D.C., 1991.
Office of Space Science and Applications, *Space Biology Plant Program* Plan, Life Sciences Division, NASA, Washington, D.C., 1991.
Office of Space Science and Applications, *Space Life Sciences Strategic Plan*, Life Sciences Division, NASA, Washington, D.C., 1992.
Office of Space Science and Applications, *Space Radiation Health Program Plan*, Life Sciences Division, NASA, Washington, D.C., 1991.
Office of Technology Assessment, *Exploring the Moon and Mars: Choices for the Nation*, OTA-ISC-502, U.S. Government Printing Office, Washington, D.C., 1991.
Space Environment Laboratory, *Solar Radiation Forecasting and Research to Support the Space Exploration Initiative*, NOAA Space Environment Laboratory, 1991.
Space Science Board, *HZE-Particle Effects in Manned Spaceflight*, National Academy of Sciences, Washington, D.C., 1973.
Space Science Board, *Life Beyond the Earth's Environment: The Biology of Living Organisms in Space*, National Academy of Sciences, Washington, D.C., 1979.
Space Science Board, *Origin and Evolution of Life—Implications for the Planets: A Scientific Strategy for the 1980s*, National Academy of Sciences, Washington, D.C., 1981.
Space Science Board, *Post-Viking Biological Investigations of Mars*, National Academy of Sciences, Washington, D.C., 1977.
Space Science Board, *Recommendations on Quarantine Policy for Mars, Jupiter, Saturn, Uranus, Neptune, and Titan*, National Academy of Sciences, Washington, D.C., 1978.
Space Science Board, *Space Science in the Twenty-First Century: Imperatives for the Decades 1995 to 2015—Life Sciences*, National Academy Press, Washington, D.C., 1988.
Space Science Board, *Strategy for Exploration of the Inner Planets: 1977-1987*, National Academy of Sciences, Washington, D.C., 1978.
Space Science Board, *A Strategy for Space Biology and Medical Science for the 1980s and 1990s*, National Academy Press, Washington, D.C., 1987.
Space Studies Board, National Research Council, *1990 Update to Strategy for the Exploration of the Inner Planets*, National Academy Press, Washington, D.C., 1990.
Space Studies Board, National Research Council, *International Cooperation for Mars Exploration and Sample Return*, National Academy Press, Washington, D.C., 1990.
Space Studies Board, National Research Council, *The Search for Life's Origins: Progress and Future Directions in Planetary Biology and Chemical Evolution*, National Academy Press, Washington, D.C., 1990.
Space Studies Board, National Research Council, *Assessment of Programs in Space Biology and Medicine 1991*, National Academy Press, Washington, D.C., 1991.
Space Studies Board, National Research Council, *Biological Contamination of Mars: Issues and Recommendations*, National Academy Press, Washington, D.C., 1992.
Space Studies Board, National Research Council, *Scientific Prerequisites for the Human Exploration of Space*, National Academy Press, Washington, D.C., 1993.
Space Studies Board, National Research Council, *Scientific Opportunities in the Human Exploration of Space*, National Academy Press, Washington, D.C., 1994.

Space Studies Board, National Research Council, *Radiation Hazards to Crews of Interplanetary Missions: Biological Issues and Research Strategies*, National Academy Press, Washington, D.C., 1996.

Space Studies Board, National Research Council, *Mars Sample Return: Issues and Recommendations*, National Academy Press, Washington, D.C., 1997.

Synthesis Group, *America at the Threshold, Report of the Synthesis Group on America's Space Exploration Initiative*, U.S. Government Printing Office, Washington, D.C., 1991.